ANIMALS OF ASIA
& AUSTRALIA

ANIMALS OF ASIA & AUSTRALIA

A VISUAL ENCYCLOPEDIA OF AMPHIBIANS, REPTILES AND MAMMALS IN THE ASIAN
AND AUSTRALIAN CONTINENTS, WITH OVER 350 ILLUSTRATIONS AND PHOTOGRAPHS

TOM JACKSON

CONSULTANT: MICHAEL CHINERY

southwater

This edition is published by Southwater, an imprint of Anness Publishing Ltd
Hermes House, 88–89 Blackfriars Road, London SE1 8HA;
tel. 020 7401 2077; fax 020 7633 9499
www.southwaterbooks.com; www.annesspublishing.com

If you like the images in this book and would like to investigate
using them for publishing, promotions or advertising, please visit
our website www.practicalpictures.com for more information.

UK agent: The Manning Partnership Ltd; tel. 01225 478444; fax 01225 478440; sales@manning-partnership.co.uk
UK distributor: Grantham Book Services Ltd; tel. 01476 541080; fax 01476 541061; orders@gbs.tbs-ltd.co.uk
North American agent/distributor: National Book Network; tel. 301 459 3366; fax 301 429 5746; www.nbnbooks.com
Australian agent/distributor: Pan Macmillan Australia; tel. 1300 135 113; fax 1300 135 103; customer.service@macmillan.com.au
New Zealand agent/distributor: David Bateman Ltd; tel. (09) 415 7664; fax (09) 415 8892

Publisher: Joanna Lorenz
Editorial Director: Judith Simons
Project Editor: Felicity Forster
Copy Editors: Gerard Cheshire, Richard Rosenfeld
and Jen Green
Illustrators: Stuart Carter, Jim Channell, John Francis,
Stephen Lings, Alan Male, Shane Marsh and Sarah Smith
Map Illustrator: Anthony Duke
Designer: Nigel Partridge
Production Controller: Claire Rae

ETHICAL TRADING POLICY

Illustrations appearing on pages 1–5: page 1 red panda; page 2 red kangaroo and joey;
page 3 short-nosed echidna; page 4t short-clawed otter; page 4b brush-tailed possum;
page 5 (top row, from left to right) leopard, thorny devil, crocodile; page 5 (bottom row,
from left to right) Bengal tiger, White's tree frog, Tasmanian devil

Previously published as part of a larger volume,
The World Encyclopedia of Animals

CONTENTS

UNDERSTANDING ANIMALS

This book examines amphibians, reptiles and mammals of Asia and Australia, but can only scratch the surface of the fantastic range of life forms packed on these continents. It concentrates on the vertebrates (the group of animals that have a backbone, which includes the above three groups as well as birds and fish); however, vertebrates form just one of the 31 major animal groups. The huge diversity of life contained in the other groups cannot rival the most familiar vertebrates for size, strength or general popularity. For most people, slugs are not in the same league as tigers, nor can crickets and crabs compete with dolphins and pandas. Yet these relatively unpopular animals are capable of amazing feats. For example, squid travel by jet propulsion, mussels change sex as they get older, and ants live in harmony in colonies of millions.

The diversity of life among the vertebrates is mind-boggling. A blue whale is 30m (100ft) long, while the smallest salamander is less than 3cm (1¼ in) long. The rest appear in every shape and size in between. Yet vertebrates are not distinguished by their bodies alone but by the unusual ways in which they use them. For example, a lizard will shed its tail if a hungry predator clamps its jaws around it, and while the tail-less reptile escapes, its attacker is kept occupied by the still-wriggling appendage. And hunting bats, which can fly thanks to the skin stretched over their elongated hand bones, so creating wings, manage to locate their prey in the dark by using sonar. They bark out high-pitched calls that bounce off surrounding objects, and the time taken for the echo to return signifies the size and location of whatever is out there. These are just two examples of the extraordinary diversity found in the vertebrate kingdom.

Left: The Bengal tiger, which can be found in India, Myanmar, Thailand, China and Indonesia is a magnificent and imposing animal, with its orange and black stripes, piercing yellow eyes and large curved claws. The species is critically endangered, their numbers having been reduced by 90 per cent over the past century due to habitat destruction and poaching.

EVOLUTION

Animals and other forms of life did not just suddenly appear on the Earth. They evolved over billions of years into countless different forms. The mechanism by which they evolved is called natural selection. The process of natural selection was first proposed by British naturalist Charles Darwin.

Many biologists estimate that there are approximately 30 million species on Earth, but to date only about two million have been discovered and recorded by scientists. So where are the rest? They live in a staggering array of habitats, from the waters of the deep oceans where sperm whales live to the deserts of Mexico, inhabited by the powerful, poisonous, gila monster lizard. The problems faced by animals in these and other habitats on Earth are very different, and so life has evolved in great variety. Each animal needs a body that can cope with its own environment.

Past evidence

At the turn of the 19th century, geologists began to realize that the world was extremely old. They used animal fossils – usually the hard remains, such as shells and bones, which are preserved in stone – to measure the age of the exposed layers of rock found in cliffs and canyons. Today we accept that the Earth is about 4.5 billion years old, but in the early 1800s the idea that the world was unimaginably old began to change people's ideas about the origins of life completely.

In addition, naturalists had always known that there was a fantastic variety of animals, but now they realized that many could be grouped into families, as if they were related. By the middle of 19th century, two British biologists had independently formulated an idea that would change the way that people saw themselves and the natural world forever. Charles Darwin and Alfred Wallace thought that the world's different animal species had gradually evolved from extinct relatives, like the ones preserved as fossils.

Darwin was the first to publish his ideas, in 1859. He had formulated them while touring South America where he studied the differences between varieties of finches and giant tortoises on the Galápagos Islands in the Pacific Ocean. Wallace came up with similar ideas about the same time, when studying different animals on the islands of South-east Asia and New Guinea.

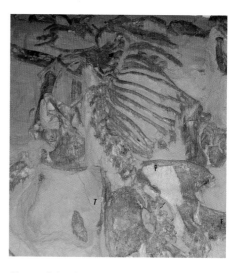

Above: Scientists know about extinct animals from studying fossils such as these mammoth bones. Fossils are the remains of dead plants or animals that have been turned to stone by natural processes over millions of years.

Survival of the fittest

Both came up with the same idea – natural selection. As breeders had known for generations, animals pass on their characteristics to their young. Darwin and Wallace suggested that wild animal species also gradually evolved through natural selection, a similar system to the artificial selection that people were using to breed prize cattle, sheep and pedigree dogs.

The theory of natural selection is often described as the survival of the fittest. This is because animals must compete with each other for limited resources including food, water, shelter and mates. But they are not all equal or exactly similar, and some members of a population of animals will have characteristics which make them "fitter" – better suited to the environment at that time.

The fitter animals will therefore be more successful at finding food and avoiding predators. Consequently, they will probably produce more offspring, many of which will also have the same characteristics as their fit parents. Because of this, the next generation

---

Jumping animals

Most animals can leap into the air, but thanks to natural selection this simple ability has been harnessed by different animals in different ways. For example, click beetles jump in somersaults to frighten off attackers, while blood-sucking fleas can leap enormous heights to move from host to host.

Above: The flying frog uses flaps of skin between its toes to glide. This allows these tree-living frogs to leap huge distances between branches.

Above: The wallaby is an excellent jumper, using its elongated hind legs to propel it in a distinctive bouncing gait. The long and powerful tail acts as a counterbalance.

will contain more individuals with the "fit" trait. And after many generations, it is possible that the whole population will carry the fit trait, since those without it die out.

Variation and time

The environment is not fixed, and does not stay the same for long. Volcanoes, diseases and gradual climate changes, for example, alter the conditions which animals have to confront. Natural selection relies on the way in which different individual animals cope with these changes. Those individuals that were once fit may later die out, as others that have a different set of characteristics become more successful in the changed environment.

Darwin did not know it, but parents pass their features on to their young through their genes. During sexual reproduction, the genes of both parents are jumbled up to produce a new individual with a unique set of characteristics. Every so often the genes mutate into a new form, and these mutations are the source of all new variations.

As the process of natural selection continues for millions of years, so groups of animals can change radically, giving rise to a new species. Life is thought to have been evolving for 3.5 billion years. In that time natural selection has produced a staggering number of species, with everything from oak trees to otters and coral to cobras.

A species is a group of organisms that can produce offspring with each other. A new species occurs once animals have changed so much that they are unable to breed with their ancestors. And if the latter no longer exist, then they have become extinct.

New species may gradually arise out of a single group of animals. In fact the original species may be replaced by one or more new species. This can happen when two separate groups of one species are kept apart by an impassable geographical feature, such as an ocean or mountain range. Kept isolated from each other, both groups then evolve in different ways and end up becoming new species.

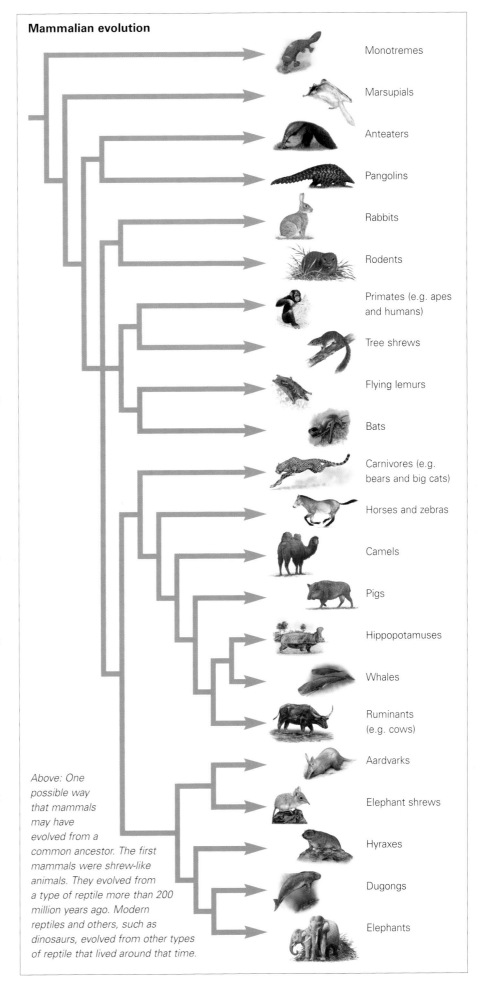

Mammalian evolution

Monotremes
Marsupials
Anteaters
Pangolins
Rabbits
Rodents
Primates (e.g. apes and humans)
Tree shrews
Flying lemurs
Bats
Carnivores (e.g. bears and big cats)
Horses and zebras
Camels
Pigs
Hippopotamuses
Whales
Ruminants (e.g. cows)
Aardvarks
Elephant shrews
Hyraxes
Dugongs
Elephants

Above: One possible way that mammals may have evolved from a common ancestor. The first mammals were shrew-like animals. They evolved from a type of reptile more than 200 million years ago. Modern reptiles and others, such as dinosaurs, evolved from other types of reptile that lived around that time.

ANATOMY

Mammals, reptiles and amphibians (which are vertebrates, as are fish and birds), come in a mind-boggling array of shapes and sizes. However all of them, from whales to bats and frogs to snakes, share a basic body plan, both inside and out.

Vertebrates are animals with a spine, generally made of bone. Bone, the hard tissues of which contain chalky substances, is also the main component of the rest of the vertebrate skeleton. The bones of the skeleton link together to form a rigid frame to protect organs and give the body its shape, while also allowing it to move. Cartilage, a softer, more flexible but tough tissue is found, for example, at the ends of bones in mobile joints, in the ears and the nose (forming the sides and the partition between the two nostrils). Some fish, including sharks and rays, have skeletons that consist entirely of cartilage.

Nerves and muscles

Vertebrates also have a spinal cord, a thick bundle of nerves extending from the brain through the spine, and down into the tail. The nerves in the spinal cord are used to control walking and other reflex movements. They also coordinate blocks of muscle that work together for an animal to move properly. A vertebrate's skeleton is on the inside, in contrast to many invertebrates, which have an outer skeleton or exoskeleton.

The vertebrate skeleton provides a solid structure which the body's muscles pull against. Muscles are blocks of protein that can contract and relax when they get an electrical impulse from a nerve.

Invertebrates

The majority of animals are invertebrates. They are a much more varied group than the vertebrates and include creatures as varied as shrimps, slugs, butterflies and starfish. Although some squid are thought to reach the size of a small whale, and while octopuses are at least as intelligent as cats and dogs, most invertebrates are much smaller and simpler animals than the vertebrates.

Below: The most successful invertebrates are the insects, including ants. This soldier army ant is defending workers as they collect food.

Reptile bodies

Reptiles have internal skeletons made from bone and cartilage. Their skins are covered in scales, which are often toughened by a waxy protein called keratin. Turtles are quite different from other reptiles. They have simpler skulls and a shell that is joined to the animal's internal skeleton.

Below: Crocodiles have very strong bodies, designed for life in and around shallow water.

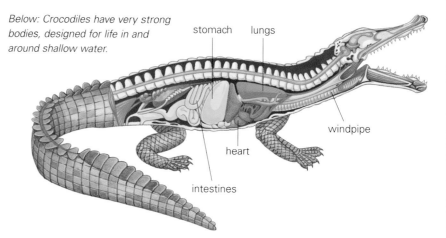

stomach   lungs   windpipe   heart   intestines

Below: Lizards have a similar body plan to crocodiles, although they are actually not very closely related.

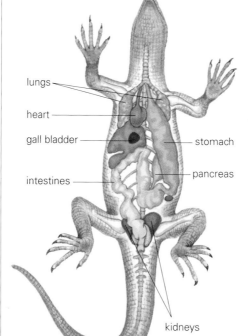

lungs   heart   gall bladder   intestines   stomach   pancreas   kidneys

Below: Snakes' internal organs are elongated so that they fit into their long, thin bodies. One of a pair of organs, such as the lungs, is often very small or missing.

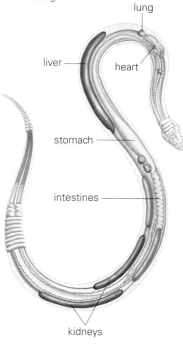

lung   liver   heart   stomach   intestines   kidneys

When on the move, the vertebrate body works like a system of pulleys, pivots and levers. The muscles are the engines of the body, and are attached to bones – the levers – by strong cables called tendons. The joint between two bones forms a pivot, and the muscles work in pairs to move a bone. For example, when an arm is bent at the elbow to raise the forearm, the bicep muscle on the front of the upper arm has to contract. This pulls the forearm up, while the tricep muscle attached to the back of the upper arm remains relaxed. To straighten the arm again, the tricep contracts and the bicep relaxes. If both muscles contract at the same time, they pull against each other, and the arm remains locked in whatever position it is in.

Vital organs

Muscles are not only attached to the skeleton. The gut – including the stomach and intestines – is surrounded by muscles. These muscles contract in rhythmic waves to push food and waste products through the body. The heart is a muscular organ made of a very strong muscle which keeps on contracting and relaxing, pumping blood around the body. The heart and other vital organs are found in the thorax, that part of the body which lies between the forelimbs. In reptiles and mammals the thorax is kept well protected inside a rib cage, as are the lungs, liver and kidneys.

Vertebrates have a single liver consisting of a number of lobes. The liver has a varied role, making chemicals required by the body and storing food. Most vertebrates also have two kidneys. Their role is to clean the blood of any impurities and toxins, and to remove excess water. The main toxins that have to be removed are compounds containing nitrogen, the by-products of eating protein. Mammal and amphibian kidneys dissolve these toxins in water to make urine. However, since many reptiles live in very dry habitats, they cannot afford to use water to remove waste, and they instead get rid of it as a solid waste similar to bird excrement.

Mammalian bodies

Most mammals are four-limbed (exceptions being sea mammals such as whales). All have at least some hair on their bodies, and females produce milk. Mammals live in a wide range of habitats and their bodies are adapted in many ways to survive. Their internal organs vary depending on where they live and what they eat.

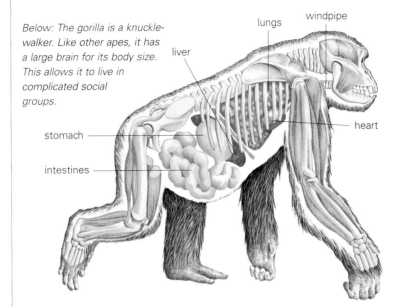

Below: The gorilla is a knuckle-walker. Like other apes, it has a large brain for its body size. This allows it to live in complicated social groups.

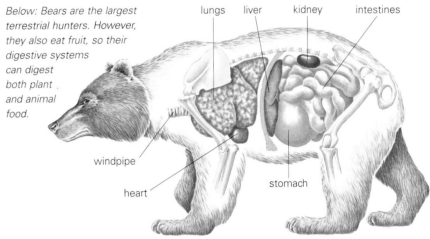

Below: Bears are the largest terrestrial hunters. However, they also eat fruit, so their digestive systems can digest both plant and animal food.

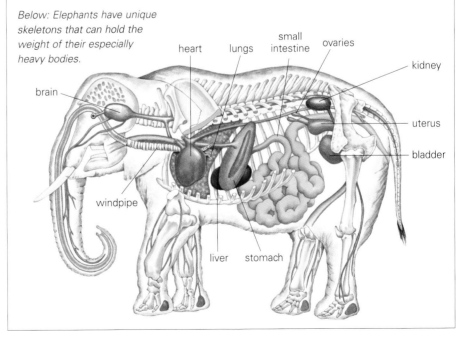

Below: Elephants have unique skeletons that can hold the weight of their especially heavy bodies.

SENSES

To stay alive, animals must find food and shelter, and defend themselves against predators. To achieve these things, they are equipped with an array of senses for monitoring their surroundings. Different species have senses adapted to nocturnal or diurnal (day-active) life.

An animal's senses are its early-warning system. They alert it to changes in its surroundings – changes which may signal an opportunity to feed or mate, or the need to escape imminent danger. The ability to act quickly and appropriately is made possible because the senses are linked to the brain by a network of nerves which send messages as electric pulses. When the brain receives the information from the senses it coordinates its response.

In many cases, generally in response to something touching the body, the signal from the sensor does not reach the brain before action is taken. Instead, it produces a reflex response which is hardwired into the nervous system. For example, when you touch a very hot object, your hand automatically recoils; you don't need to think about it.

All animals have to be sensitive to their environment to survive. Even the simplest animals, such as jellyfish and roundworms, react to changes in their surroundings. Simple animals, however, have only a limited ability to move or defend themselves, and therefore generally have limited senses. Larger animals, such as vertebrates,

have a much more complex array of sense organs. Most vertebrates can hear, see, smell, taste and touch.

Vision

Invertebrates' eyes are generally designed to detect motion. Vertebrates' eyes, however, are better at forming clear images, often in colour. Vertebrates' eyes are balls of clear jelly which have an inner lining of light-sensitive cells. This lining, called the retina, is made up of one or two types of cell. The rod cells – named after their shape – are very sensitive to all types of light, but are only capable of forming black and white images. Animals which are active at night generally have (and need) only rods in their eyes.

Colour vision is important for just a few animals, such as monkeys, which need, for example, to see the brightest and therefore ripest fruits. Colour images are made by the cone cells – so named because of their shape – in the retina. There are three types of cone, each of which is sensitive to a particular wavelength of light. Low wavelengths appear as reds, high wavelengths as blues, with green colours being detected in between.

Above: Frogs have large eyes positioned on the upper side of the head so that the animals can lie mainly submerged in water with just their eyes poking out.

The light is focused on the retina by a lens to produce a clear image. Muscles change the shape of the lens so that it can focus the light arriving from different distances. While invertebrates may have several eyes, all vertebrates have just two, and they are always positioned on the head. Animals such as rabbits, which are constantly looking out for danger, have eyes on the side of the head to give a wide field of vision. But while they can see in almost all directions, rabbits have difficulty judging distances and speeds. Animals that have eyes pointing forward are better at doing this because each eye's field of vision overlaps with the other. This binocular vision helps hunting animals and others, such as tree-living primates, to judge distances more accurately.

Eyes can also detect radiation in a small band of wavelengths, and some animals detect radiation that is invisible to our eyes. Flying insects and birds can see ultraviolet light, which extends the range of their colour vision. At the other end of the spectrum many snakes can detect radiation with a lower wavelength. They sense infrared, or heat, through pits on the face which enables them to track their warm-blooded prey in pitch darkness.

Below: This slow loris is nocturnal, that is, night-active. It has very large eyes that collect as much light as possible so that it can see in the gloom of the night.

Below: Like other hunters, a seal has eyes positioned on the front of its head. Forward-looking eyes are useful for judging distances, making it easier to chase down prey.

Hearing

An animal's brain interprets waves of pressure travelling through the air, and detected by the ears, as sound. Many animals do not hear these waves with ears but detect them in other ways instead. For example, although snakes can hear, they are much more sensitive to vibrations through the lower jaw, travelling through the ground. Long facial whiskers sported by many mammals, from cats to dugongs, are very sensitive touch receptors. They can be so sensitive that they will even respond to currents in the air.

In many ways, hearing is a sensitive extension of the sense of touch. The ears of amphibians, reptiles and mammals have an eardrum which is sensitive to tiny changes in pressure. An eardrum is a thin membrane of skin which vibrates as the air waves hit it. A tiny bone (or in the case of mammals, three bones) attached to the drum transmit the vibrations to a shell-shaped structure called a cochlea. The cochlea is filled with a liquid which picks up the vibrations. As the liquid moves inside the cochlea, tiny hair-like structures lining it wave back and forth. Nerves stimulated by this wave motion send the information to the brain, which interprets it as sound.

A mammal's ear is divided into three sections. The cochlea forms the inner ear and the middle ear consists of the bones between the cochlea and eardrum. The outer ear is the tube joining the outside world and the

auricle – the fleshy structure on the side of the head that collects the sound waves – to the middle ear. Amphibians and reptiles do not possess auricles. Instead their eardrums are either on the side of the head – easily visible on many frogs and lizards – or under the skin, as in snakes.

Smell and taste

Smell and taste are so closely related as to form a single sense. Snakes and lizards, for example, taste the air with their forked tongues. However, it is perhaps the most complex sense. Noses, tongues and other smelling

Above: Snakes have forked tongues that they use to taste the air. The tips of the fork are slotted into an organ in the roof of the mouth. This organ is linked to the nose, and chemicals picked up by the tongue are identified with great sensitivity.

organs are lined with sensitive cells which can analyze a huge range of chemicals that float in the air or exist in food. Animals such as dogs, which rely on their sense of smell, have long noses packed with odour-sensitive cells. Monkeys, of the other hand, are less reliant on a sense of smell, and consequently have short noses capable only of detecting stronger odours.

Below: Hares have very large outer ears which they use like satellite dishes to pick up sound waves. They can rotate each ear separately to detect sound from all directions.

Below: Lizards do not have outer ears at all. Their hearing organs are contained inside the head and joined to the outside world through an eardrum membrane.

Below: Monkeys rely on their eyesight more than their sense of smell, so their noses tend to be short and small compared to those of dogs, rabbits and rats.

SURVIVAL

In order to stay alive, animals must not only find enough food, but also avoid becoming a predator's meal. To achieve this, animals have evolved many strategies to feed on a wide range of foods, and an array of weapons and defensive tactics to keep safe.

An animal must keep feeding in order to replace the energy used in staying alive. Substances in the food, such as sugars, are burned by the body, and the subsequent release of energy is used to heat the body and power its movements. Food is also essential for growth. Although most growth takes place during the early period of an animal's life, it never really stops because injuries need to heal and worn-out tissues need replacing. Some animals continue growing throughout life. Proteins in the food are the main building blocks of living bodies.

Plant food

Some animals will eat just about anything, while others are much more fussy. As a group, vertebrates get their energy from a wide range of sources – everything from shellfish and wood to honey and blood. Animals are often classified according to how they feed, forming several large groups filled with many otherwise unrelated animals.

Animals that eat plants are generally grouped together as herbivores. But this term is not very descriptive because there is such a wide range of plant foods. Animals that eat grass are known as grazers. However, this term can also apply to any animal which eats any plant that covers the ground

Above: The thorny devil is well adapted to life in the arid desert: when rain or dew lands on its back, the water flows along grooves on its back, leading to the corners of its mouth, where it can then drink the moisture.

in large amounts, such as seaweed or sedge. Typical grazers include bison and wildebeest but some, such as the marine iguana or gelada baboon, are not so typical. Animals such as giraffes or antelopes, which pick off the tastiest leaves, buds and fruit from bushes and trees, are called browsers. Other browsing animals include many monkeys, but some monkeys eat only leaves (the folivores) or fruit (the frugivores).

Many monkeys have a much broader diet, eating everything from insects to the sap which seeps out from the bark of tropical trees. Animals that eat both plant and animal foods are called omnivores. Bears are omnivorous, as are humans, but the most catholic of tastes belong to scavenging animals, such as rats and other rodents, which eat anything they can get their teeth into. Omnivores in general, and scavengers in particular, are very curious animals. They will investigate anything that looks or smells like food, and if it also tastes like food, then it probably is.

A taste for flesh

The term carnivore is often applied to any animal that eats flesh, but it is more correctly used to refer to an order of mammals which includes cats, dogs, bears and many smaller animals, such as weasels and mongooses. These animals are the kings of killing, armed with razor-sharp claws and powerful jaws crammed full of chisel-like teeth. They use their strength and speed to overpower their prey, either by running them down or taking them by surprise with an ambush.

Below: Going a step further than most herbivores, the giant panda only eats one type of vegetation: bamboo. Its large molar teeth help it to break down the tough stems.

Below: The larger a crocodile grows, the wider the variety of food it eats, although even large individuals feed mainly on small prey such as fish, frogs, crabs and turtles.

*Above: The leopard is an opportunistic
predator, hunting several different kinds of
prey, for example rodents, birds, hares and
deer. They often hide their food in trees where
most other predators can't reach.*

However, land-dwelling carnivores
are not the only expert killers. The
largest meat-eater is the orca, or killer
whale, which is at least three times
the size of the brown bear, the largest
killer on land.

While snakes are much smaller in
comparison, they are just as deadly, if
not more so. They kill in one of two
ways, either suffocating their prey by
wrapping their coils tightly around
them, or by injecting them with a
poison through their fangs.

Arms race
Ironically, the same weapons used by
predators are often used by their prey
to defend themselves. For example,
several species of frog, toad and
salamander secrete poisons on to their
skin. In some cases, such as the poison-
dart frog, this poison is enough to kill
any predator that tries to eat it, thus
making sure that the killer won't
repeat its performance. More often,
though, a predator finds that its meal
tastes horrible and remembers not
to eat another one again. To remind
the predators to keep away, many
poisonous amphibians are brightly

coloured, which ensures that they are
easily recognized.

Many predators rely on stealth to
catch their prey, and staying hidden is
part of the plan. A camouflaged coat,
such as a tiger's stripes, helps animals
blend into their surroundings. Many
species also use this technique to
ensure that they do not get eaten.
Most freeze when danger approaches,
and then scurry to safety as quickly
as possible. Chameleons have
taken camouflage to an even more
sophisticated level as they can change
the colour of their scaly skins, which
helps them to blend in with their
surrounding environment.

Plant-eating animals that live in the
open cannot hide from predators that
are armed with sharp teeth and claws.
And the plant-eaters cannot rely on
similar weapons to defend themselves.
They are outgunned

*Right: Chameleons
have skin cells that
can be opened and
closed to make
their skin colour
change.*

because they do not possess sharp,
pointed teeth but flattened ones to
grind up their plant food. The best
chance they have of avoiding danger is
to run away. Animals such as antelopes
or deer consequently have long, hoofed
feet that lengthen their legs considerably;
they are, in fact, standing on their
toenails. These long legs allow them
to run faster and leap high into the
air to escape an attacker's jaws.

Animals that do not flee must stand
and fight. Most large herbivores are
armed with horns or antlers. Although
used chiefly for display, the horns are
the last line of defence when cornered.

REPRODUCTION

All animals share the urge to produce offspring which will survive after the parents die. The process of heredity is determined by genes, through which characteristics are passed from parents to offspring. Reproduction presents several problems, and animals have adopted different strategies for tackling them.

Animals have two main goals: to find food and a mate. To achieve these goals, they must survive everything that the environment throws at them, from extremes of the weather, such as floods and droughts, to hungry predators. They have to find sufficient supplies of food, and on top of that locate a mate before their competitors. If they find sufficient food but fail to produce any offspring, their struggle for survival will have been wasted.

One parent or two?

There are two ways in which an animal can reproduce, asexually or sexually. Animals that are produced by asexual reproduction, or parthenogenesis, have only one parent, a mother. The offspring are identical to their mother and to each other. Sexual reproduction involves two parents of the opposite sex. The offspring are hybrids of the two parents, with a mixture of their parents' characteristics.

The offspring inherit their parents' traits through their genes. Genes can be defined in various ways. One simple definition is that they are the unit of inheritance – a single inherited

Below: Crocodiles bury their eggs in a nest. The temperature of the nest determines the sex of the young reptiles. Hot nests produce more males than cool ones. Crocodile mothers are very gentle when it comes to raising young.

Above: Many male frogs croak by pumping air into an expandable throat sac. The croak is intended to attract females. The deeper the croak, the more attractive it is. However, some males lurk silently and mate with females as they approach the croaking males.

Above: In deer and many other grazing animals, the males fight each other for the right to mate with the females in the herd. The deer with the largest antlers often wins without fighting, and real fights only break out if two males appear equally well-endowed.

characteristic which cannot be subdivided any further. Genes are also segments of DNA (deoxyribonucleic acid), a complex chemical that forms long chains. It is found at the heart of every living cell. Each link in the DNA chain forms part of a code that controls how an animal's body develops and survives. And every cell in the body contains a full set of DNA which could be used to build a whole new body.

Animals produced through sexual reproduction receive half their DNA, or half their genes, from each parent. The male parent provides half the supply of genes, contained in a sperm. Each sperm's only role is to find its way to, and fertilize, an egg, its female equivalent. Besides containing the other half of the DNA, the egg also holds a supply of food for the offspring as it develops into a new individual. Animals created through parthenogenesis get all their genes from their mother, and all of them are therefore the same sex – female.

Pros and cons

All mammals reproduce sexually, as do most reptiles and amphibians. However, there are a substantial number of reptiles and amphibians, especially lizards, which reproduce by parthenogenesis. There are benefits and disadvantages to both types of reproduction. Parthenogenesis is quick and convenient. The mother does not need to find a mate, and can devote all of her energy to producing huge numbers of young. This strategy is ideal for populating as yet unexploited territory. However, being identical, these animals are very vulnerable to attack. If, for example, one is killed by a disease or outwitted by a predator, it is very likely that they will all suffer the same fate. Consequently, whole communities of animals produced through parthenogenesis can be wiped out.

Sexual animals, on the other hand, are much more varied. Each one is unique, formed by a mixture of genes from both parents. This variation means that a group of animals produced by sexual reproduction is more likely to triumph over adversity than a group of asexual ones. However, sexual reproduction takes up a great deal of time and effort.

Attracting mates

Since females produce only a limited number of eggs, they are keen to make sure that they are fertilized by a male with good genes. If a male is fit and healthy, this is a sign that he has good genes. Good genes will ensure that the offspring will be able to compete with other animals for food and mates of their own. Because the females have the final say in agreeing to mate, the

Above: Once an adult male orang-utan develops its large cheek flaps, it is able to attract sexually receptive females. Because they are solitary animals, there is a strong bond between females and their young.

Below: The female kangaroo is able to produce two types of milk simultaneously: fat-rich milk for a larger joey and carbohydrate-rich milk for a smaller pouch-bound sibling.

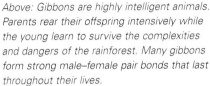

Above: Gibbons are highly intelligent animals. Parents rear their offspring intensively while the young learn to survive the complexities and dangers of the rainforest. Many gibbons form strong male–female pair bonds that last throughout their lives.

males have to put a lot of effort into getting noticed. Many are brightly coloured, make loud noises, and they are often larger than the females. In many species the males even compete with each other for the right to display to the females. Winning that right is a good sign that they have the best genes.

Parental care

The amount of care that the offspring receive from their parents varies considerably. There is a necessary trade-off between the amount of useful care parents can give to each offspring, the number of offspring they can produce and how regularly they can breed. Mammals invest heavily in parental care, suckling their young after giving birth, while most young amphibians or reptiles never meet their parents at all.

By suckling, mammals ensure that their young grow to a size where they can look after themselves. Generally, the young stay with the mother until it is time for her to give birth to the next litter – at least one or two months. However, in many species, including humans, the young stay with their parents for many years.

Other types of animals pursue the opposite strategy, producing large numbers of young that are left to fend for themselves. The vast majority in each batch of eggs – consisting of hundreds or even thousands – die before reaching adulthood, and many never even hatch. The survival rates, for example of frogs, are very low.

Animals that live in complicated societies, such as elephants, apes and humans, tend to produce a single offspring every few years. The parents direct their energies into protecting and rearing the young, giving them a good chance of survival. Animals which live for a only a short time, such as mice, rabbits, and reptiles and amphibians in general, need to reproduce quickly to make the most of their short lives. They produce high numbers of young, and do not waste time on anything more than the bare minimum of parental care. If successful, these animals can reproduce at an alarming pace.

AMPHIBIANS

Amphibians are the link between fish and land animals. One in eight of all vertebrate animals are amphibians. This group includes frogs, toads and newts as well as rarer types, such as giant sirens, hellbenders and worm-like caecilians. Amphibians are equally at home in water and on land.

Amphibians live on every continent except for Antarctica. None can survive in salt water, although a few species live close to the sea in the brackish water at river mouths. Being cold-blooded – their body temperature is always about the same as the temperature of their surroundings – most amphibians are found in the warmer regions of the world.

Unlike other land vertebrates, amphibians spend the early part of their lives in a different form from that of the adults. As they grow, the young gradually metamorphose into the adult body. Having a larval form means that the adults and their offspring live and feed in different places. In general the larvae are aquatic, while the adults spend most of their time on land.

The adults are hunters, feeding on other animals, while the young are generally plant eaters, filtering tiny plants from the water or grazing on aquatic plants which line the bottom of ponds and rivers.

Below: Amphibians must lay their eggs near a source of water. In most cases, such as this frog spawn, the eggs are laid straight into a pond or swamp. The tadpoles develop inside the jelly-like egg and then hatch out after the food supply in the egg's yolk runs out.

Life changing

Most amphibians hatch from eggs laid in water or, in a few cases, in moist soil or nests made of hardened mucus. Once hatched, the young amphibians, or larvae, live as completely aquatic animals. Those born on land wriggle to the nearest pool of water or drop from their nest into a river.

The larvae of frogs and toads are called tadpoles. Like the young of salamanders – a group that includes all other amphibians except caecilians – tadpoles do not have any legs at first. They swim using a long tail that has a fish-like fin extending along its length. As they grow, the larvae sprout legs. In general the back legs begin to grow first, followed by the front pair. Adult frogs do not have tails, and after the first few months a tadpole's tail begins to be reabsorbed into the body – it does not just fall away.

All adult salamanders keep their tails, and those species that spend their entire lives in water often retain the fin along the tail, along with other parts, such as external gills, a feature that is more commonly seen in the larval stage.

Above: Amphibians begin life looking very different from the adult form. Most of the time these larval forms, such as this frog tadpole, live in water as they slowly develop into the adult form, growing legs and lungs so that they can survive on land.

Body form

Amphibian larvae hatch with external gills but, as they grow, many (including all frogs and the many salamanders which live on land) develop internal gills. In most land-living species these internal gills are eventually replaced by lungs. Amphibians are also able to absorb oxygen directly through the skin, especially through the thin and moist tissues inside the mouth. A large number of land-living salamanders get all their oxygen in this way because they do not have lungs.

All adult frogs and toads return to the water to breed and lay their eggs, which are often deposited in a jelly-like mass called frog spawn. Several types of salamander do not lay eggs, and instead the females keep the fertilized eggs inside their bodies. The larvae develop inside the eggs, fed by a supply of rich yoke, and do not hatch until they have reached adult form.

Above: After the first few weeks, a tadpole acquires tiny back legs. As the legs grow, the long tail is gradually reabsorbed into the body. The front legs appear after the back ones have formed.

Adult form

Most adult amphibians have four limbs, with four digits on the front pair and five on the rear. Unlike other land-living animals, such as reptiles or mammals, their skin is naked and soft. Frogs' skin is smooth and moist, while toads generally have a warty appearance.

The skins of many salamanders are brightly coloured, with patterns that often change throughout the year. Colour change prior to the mating season signals the salamander's readiness to mate. Many frogs also have bright skin colours. Although their skin shades can change considerably in different light levels, these colours are generally not mating signals to fellow frogs. Instead they are warnings to predators that the frog's skin is laced with deadly poison. While toads tend to be drab in colour, many also secrete toxic chemicals to keep predators away. These substances are often stored in swollen warts which burst when the toad is attacked.

Below: Adult frogs may live in water or on land. Aquatic ones have webbed feet, while those on land have powerful legs for jumping and climbing. All frogs must return to a source of water to mate.

Forever young

Salamanders which have changed into adults, but which have not yet reached adult size, are called efts. The time it takes for an amphibian to grow from a newly hatched larva to an adult varies considerably, and the chief factor is the temperature of the water in which it is developing. Most frogs and toads develop in shallow waters, warmed by the summer sun, and they generally reach adulthood within three to four months. However, salamanders, especially the largest ones, can take much longer, and at the northern and southern limits of their geographical spread some salamanders stay as larvae for many years. It appears that the trigger for the change into adult form is linked to the temperature, and in cold climates this change happens only every few years. In fact it may not happen during a salamander's lifetime, and consequently several species have evolved the ability to develop sexual organs even when they still look like larvae.

Below: Marbled salamanders are unusual in that the females lay their eggs on dry land and coil themselves around them to keep them as moist as possible. They stay like this until the seasonal rains fall. The water stimulates the eggs to hatch.

REPTILES

Reptiles include lizards, snakes, alligators, crocodiles, turtles and tortoises, as well as now-extinct creatures such as dinosaurs and the ancestors of birds. Crocodiles have roamed the Earth for 200 million years and are still highly successful hunters.

Reptiles are a large and diverse group of animals containing about 6,500 species. Many of these animals look very different from each other and live in a large number of habitats, from the deep ocean to the scorching desert. Despite their great diversity, all reptiles share certain characteristics.

Most reptiles lay eggs, but these are different from those of an amphibian because they have a hard, thin shell rather than a soft, jelly-like one. This protects the developing young inside and, more importantly, stops them from drying out. Shelled eggs were an evolutionary breakthrough because they meant that adult reptiles did not have to return to the water to breed. Their waterproof eggs could be laid anywhere, even in the driest places. Reptiles were also the first group of land-living animals to develop into an adult form inside the egg. They did not emerge as under-developed larvae like the young of most amphibians.

Below: Alligators and other crocodilians are an ancient group of reptiles that have no close living relatives. They are archosaurs, a group of reptiles that included the dinosaurs. Other living reptiles belong to a different group.

Released from their ties to water, the reptiles developed unique ways of retaining moisture. Their skins are covered by hardened plates or scales to stop water being lost. The scales are also coated with the protein keratin, the same substance used to make fingernails and hair.

All reptiles breathe using lungs; if they were to absorb air through the skin it would involve too much water loss. Like amphibians, reptiles are cold-blooded and cannot heat their bodies from within as mammals can. Consequently, reptiles are commonly found in warm climates.

Ancient killers

Being such a diverse group, reptiles share few defining characteristics besides their shelled eggs, scaly skin and lungs. They broadly divide into four orders. The first contains the crocodiles, and includes alligators and caimans; these are contemporaries of the dinosaurs, both groups being related to a common ancestor.

In fact today's crocodiles have changed little since the age when dinosaurs ruled the world over 200 million years

Above: Turtles and their relatives, such as these giant tortoises, are unusual reptiles. Not only do they have bony shells fused around their bodies, but they also have skulls that are quite different from other reptiles. Turtles are also unusual because many of them live in the ocean, while most reptiles live on land.

ago. Unlike the dinosaurs, which disappeared 65 million years ago, the crocodiles are still going strong. Technically speaking, the dinosaurs never actually died out; their direct descendents, the birds, are still thriving. Although birds are now grouped separately from reptiles, scientists know that they all evolved from ancestors which lived about 400 million years ago. Mammals, on the other hand, broke away from this group about 300 million years ago.

Above: Most reptiles, including this tree boa, lay eggs. The young hatch looking like small versions of the adults. However, several snakes and lizards give birth to live young, which emerge from their mother fully formed.

Distant relatives

The second reptile order includes turtles, terrapins and tortoises. These are only distantly related to other reptiles, and it shows. Turtles are also the oldest group of reptiles, evolving their bony shells and clumsy bodies before crocodiles, dinosaurs or any other living reptile group existed. Although turtles evolved on land, many have since returned to water. However, they still breathe air, and all must return to land to lay their eggs.

The third group of reptiles is the largest. Collectively called the squamates, this group includes snakes and lizards.

Snakes, with their legless bodies and formidable reputations, are perhaps the most familiar reptiles. They evolved from animals that did have legs, and many retain tiny vestiges of legs. The squamates include other legless members such as the amphisbaenians (or worm lizards) and slow worms. Both of these groups are more closely related to lizards than snakes, despite looking more like the latter. Lizards are not a simple group of reptiles, and many biologists refer to them as several different groups, including the skinks, monitors, geckos and iguanas.

Below: Lizards, such as this iguana, are the largest group of reptiles. Most are hunters that live in hot parts of the world, and they are especially successful in dry areas where other types of animal are not so common.

The squamates are so diverse in their lifestyles and body forms that it is hard to find factors which they have in common. One feature not found in other reptile orders is the Jacobson's organ. It is positioned in the roof of the mouth and is closely associated with the nose. All snakes and most lizards use this organ to "taste" the air and detect prey. The long forked tongue of most of these animals flicks out into the air, picking up tiny particles on its moist surface. Once back inside the mouth, each fork slots into the Jacobson's organ which then analyzes the substances.

The fourth and final order of reptiles is very small: the tuataras. These include just a few species, all of which are very rare indeed, clinging to life on islands off the mainland of New Zealand. To most people a tuatura looks like a large iguana. However, scientists believe that it is only a distant relative of lizards and other squamates because it has an odd-shaped skull and no eardrums.

Tuatara

Despite their differences from lizards and other squamates, tuataras do share one feature with lizards: the so-called third eye. This light-sensitive gland inside the head can detect light penetrating the thin skull. Both types of reptile use the third eye to regulate their exposure to the sun and so regulate their body temperature throughout the year.

Below: The tuatara is a living fossil, living on just a few islands around New Zealand. It looks like a lizard, but its skull and skeleton show that it is the last member of another ancient group of reptiles.

PLACENTAL MAMMALS

Mammals are the most familiar of all vertebrates. This is because not only are human beings mammals, but also most domestic animals and pets belong in this category. Placental mammals are also more widespread than other types of animal, being found in all parts of the world.

Mammals are grouped together because they share a number of characteristics. However, these common features do not come close to describing the huge diversity within the mammal class. For example, the largest animal that has ever existed on Earth – the blue whale – is a mammal, and so this monster of the deep shares several crucial traits with even the smallest mammals, such as the tiniest of shrews. Other mammals include elephants and moles, monkeys and hippopotamuses, and bats and camels. To add to this great diversity, mammals live in more places on Earth than any other group of animals, from the frozen ice fields of the Arctic to the humid treetops of the Amazon rainforest, and even under the sandy soil of African deserts.

Mammal bodies

The most obvious mammalian feature is hair. All mammals have hair made of keratin protein and, in most cases, it forms a thick coat of fur, though many

Below: Although they are often mistaken for fish, dolphins are mammals: they breathe air and suckle their young. However, life under water requires flippers and fins, not legs.

Above: Tigers are the largest of the cat family, having powerful jaws and teeth, sharp claws and strong forelegs. Like many mammals, they rely on water to cool themselves down, particularly in the tropics. Cubs are reliant on their mother's milk for the first eight weeks.

mammals are relatively naked, not least humans. Unlike reptiles and amphibians, all mammals are warm-blooded, which means that they can keep their body temperature at a constant level. While this requires quite a lot of energy, it means that mammals are not totally dependent on the temperature of their surroundings. In places where other vertebrates would be frozen solid, mammals can survive by seeking out food and keeping themselves warm. Many mammals, including humans, can also cool their bodies using sweat. The water secreted on the skin cools the body efficiently, but it does mean that these animals need to drink more replacement water than do other groups.

Incidentally, the name mammal comes from the mammary glands. These glands are the means by which all female mammals provide milk (or liquid food) to their developing young. The young suck the milk through teats or nipples for the first few weeks or months of life.

Reproduction

Mammals reproduce in a number of ways. Monotremes, such as the duck-billed platypus, lay eggs, but all other mammals give birth to their young. Marsupials, a relatively small group of animals which includes kangaroos, give birth to very undeveloped young which then continue to grow inside a fold or pouch on the mother's skin.

The majority of mammals, called the placental mammals or eutherians, do not give birth to their young until they are fully formed and resemble the adults. The developing young, or foetuses, grow inside a womb or uterus where they are fed by the mother through a placenta. This large organ allows the young to stay inside the mother for a lot longer than in most other animals. It forms the interface between the mother's blood supply and that of the developing foetus, where oxygen and food pass from the parent to her offspring. The placenta is attached to the foetus by means of an umbilical cord which withers and drops off soon after the birth.

Widespread range

Mammals are found in a wider variety of habitats than any other group of animals. While mammals all breathe air with their

Right: One factor that makes mammals unique is that the females have mammary glands. These glands produce milk for the young animals to drink, as this fallow deer fawn is doing. The milk is a mixture of fat, protein and sugars.

Above: The sun bear is the smallest species of bear, and like other bears its cubs are born blind and hairless, totally dependent on their mother for survival for the first two years.

lungs, this has not prevented many from making their homes in water. In many ways the streamlined bodies of whales and dolphins, for example, resemble those of sharks and other large fish. However, they are very much mammals, breathing air through a large nostril or blowhole in the top of the head, but their body hair has been reduced to just a few thick bristles.

Above: Plenty of mammals can glide, but only bats join birds and insects in true flight. A bat wing is made from skin that is stretched between long finger bones.

At the other end of the spectrum, some mammals even fly. Bats darting through the gloom of a summer evening may appear to be small birds, but they too are mammals with furry bodies and wings made from stretched skin instead of feathers. Although most other mammals have a more conventional body plan, with four legs and a tail, they too have evolved to survive in a startling range of habitats. They have achieved this not just by adapting their bodies but by changing their behaviour. In general, mammals have larger brains than reptiles and amphibians, and this allows them to understand their environment more fully. Many mammals, such as monkeys and dogs, survive by living in complex social groups in which individuals cooperate with each other when hunting food, protecting the group from danger and even finding mates.

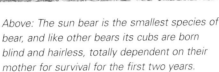

MONOTREMES AND MARSUPIALS

Not all mammals begin their lives as foetuses growing inside a uterus within their mother. Monotremes and marsupials develop in other ways. Monotremes are the only mammals to lay eggs like birds or lizards, while marsupials raise their young inside pouches on the outside of their bodies.

Egg-laying mammals

Monotremes are a group of just three mammal species which still lay their young inside eggs: two species of echidnas, short and long-nosed, (also called spiny anteaters), and the odd-looking duck-billed platypus. Although monotremes are hairy and feed their young with milk, they are distant relatives of other mammals.

Monotremes lay two or three very tiny spherical eggs which have a much softer shell than those of birds or many reptiles. And unlike other mammals, monotremes do not have a birth canal. Instead, their eggs travel through the same body opening as the urine and faeces. A single multi-purpose body opening like this is called a cloaca (Latin for drain), and is a feature that monotremes share with birds and reptiles. In fact, the name monotreme means one-holed animal.

The young of monotremes stay in the egg for only a matter of days before hatching out, and then they continue their development while being nursed by their mother. Nursing

Above: Monotremes, such as this echidna, are in the minority compared to other mammals. They lay eggs, more like a bird or a reptile, rather than give birth to their young. However, the monotremes are mammals because they feed their young on a supply of milk.

echidnas keep their young in a pouch on their underside, while platypus young spend their early days in an underground nest. While all other mammals have teats for delivering milk, the monotremes do not possess this adaptation. Instead, the mother secretes the fatty liquid on her fur and the young then lap it up.

Monotremes hatch out in a very undeveloped state. They are hairless, blind, barely 1cm (⅓in) long and have just blunt buds for limbs. The young are dependent on their mothers' milk for a long time – six months in the case of echidnas. Although they are a tiny group compared to other mammal orders, the monotremes are highly adapted to their habitats and are quite amazing mammals.

Life in a pouch

Although their 270 species make the marsupials a much larger group than the monotremes, they are still very much the minority compared to the

Below: Kangaroos are the most familiar marsupial mammals. Like most marsupials, the young grow inside a pouch on the mother's belly, where they are fed on milk from a nipple. Some other marsupials do not have such a developed pouch.

Monotreme features

Monotremes were living nearly 200 million years ago, long before any other types of mammal existed. However, this does not mean that animals similar to monotremes were the ancestors of all modern mammals. Further, today's monotremes (which live only in Australasia) are by no means primitive. In fact they have some highly developed features which are not found in any other mammal group. For example, male platypuses sport a venomous spur behind the knee which they use when defending themselves and fighting each other.

Below: Because of its strange appearance, the duck-billed platypus was thought to be a fake when specimens were first displayed. However, it is a very well-adapted underwater hunter, using its bill to detect the electric fields produced by its prey's muscles.

Above: Most marsupials live in Australia where there are few large predators. The koala is adapted to living in trees and eating leaves, and has evolved features that are seen in other animals who live in a similar way, for example grasping hands and a long, fermenting digestive system.

4,000 or so species of placental mammals that inhabit every corner of the world. Marsupials are found only in Australia and the islands of New Guinea, and a few species live in the Americas.

Marsupials are much more closely related to placental mammals than monotremes. Biologists believe that marsupials evolved first in the Americas and then spread to Australia via Antarctica, in the days when the three continents were joined together. Once Australia became an island, its marsupials remained the dominant mammals, while placental mammals began to take over from them elsewhere in the world.

The placental mammals have a single uterus connected to both ovaries by oviducts, while marsupials have two much smaller uteruses, each connected to a single ovary. The uterus of a placental mammal can swell up as the foetus grows. However, those of a

marsupial cannot, which means that the young have to be born much earlier.

Like a newly hatched monotreme, the tiny marsupial baby is hairless and blind, and has stubby forelegs. After two or three weeks in the uterus, the baby makes its way out through a birth canal that grows especially for this journey. Once outside, the baby battles through its mother's fur to her nipples.

Many marsupials have their nipples inside a pouch or marsupium. Once inside, the baby latches on to

Right: Tasmanian devils used to be found right across Australia. Now they only survive on the southern island of Tasmania. Many other marsupial species have suffered due to the activities of humans and their domestic animals.

the nipple which swells inside its mouth to create a very firm connection. The larger marsupials, such as red kangaroos, spend several months inside the pouch and may later be joined by younger siblings. The mother's milk supply is tailored for each of her young, with each baby getting the right amount of fats and proteins for its age. A baby marsupial that is old enough to leave the pouch is called a joey. Joeys often return to their mother's pouch to sleep or feed for several months more before becoming completely independent.

ECOLOGY

Ecology is the study of how groups of organisms interact with members of their own species, other organisms and the environment. All types of animals live in a community of interdependent organisms called an ecosystem, in which they have their own particular role.

The natural world is filled with a wealth of opportunities for animals to feed and breed. Every animal species has evolved to take advantage of a certain set of these opportunities, called a niche. A niche is not just a physical place but also a lifestyle exploited by that single species. For example, even though they live in the same rainforest habitat, sloths and tapirs occupy very different niches.

To understand how different organisms interrelate, ecologists combine all the niches in an area into a community, called an ecosystem. Ecosystems do not really exist because it is impossible to know where one ends and another begins, but the system is a useful tool when learning more about the natural world.

Food chains

One way of understanding how an ecosystem works is to follow the food chains within it. A food chain is made up of a series of organisms that prey on each other. Each habitat is filled with them, and since they often merge into and converge from each other, they are often combined into food webs.

Below: Frogs are often in the middle of a food chain: plants are eaten by grasshoppers, grasshoppers are eaten by frogs, and frogs are eaten by birds of prey. For this reason, there tends to be more frogs than birds.

Ecologists use food chains to see how energy and nutrients flow through natural communities. Food chains always begin with plants. Plants are the only organisms on Earth that do not need to feed, deriving their energy from sunlight, whereas all other organisms, including animals, get theirs from food. At the next level up the food chain come the plant-eaters. They eat the plants, and extract the sugar and other useful substances made by them. And, like the plants, they use these substances to power their bodies and stay alive. The predators occupy the next level up, and they eat the bodies of the plant-eating animals.

At each stage of the food chain, energy is lost, mainly as heat given out by the animals' bodies. Because of this, less energy is available at each level up the food chain. This means that in a healthy ecosystem there are always fewer predators than prey, and always more plants than plant-eaters.

Nutrient cycles

A very simple food chain would be as follows: grass, wildebeest and lion. However, the reality of most ecosystems is much more complex, with many more layers, including certain animals that eat both plants and animals. Every food chain ends with a top predator, in our example, the lion. Nothing preys on the lion, at least when it is alive, but once it dies the food chain continues as insects, fungi and other decomposers feed on the carcass. Eventually nothing is left of the lion's body. All the energy stored in it is removed by the decomposers, and the chemicals which made up its body have returned to the environment as carbon dioxide gas, water and minerals in the soil. And these are the very same substances needed by a growing plant. The cycle is complete.

Above: Nothing is wasted in nature. The dung beetle uses the droppings of larger grazing animals as a supply of food for its developing young. Since the beetles clear away all the dung, the soil is not damaged by it, the grass continues to grow, and the grazers have plenty of food.

Living together

As food chains show, the lives of different animals in an ecosystem are closely related. If all the plants died for some reason, it would not just be the plant-eaters that would go hungry. As all of them begin to die, the predators would starve too. Only the decomposers might appear to benefit. Put another way, the other species living alongside an animal are just as integral to that animal's environment as the weather and landscape. This is yet another way of saying that animal species have not evolved isolated from each another.

The result is that as predators have evolved new ways of catching their prey, the prey has had to evolve new ways of escaping. On many occasions this process of co-evolution has created symbiotic relationships between two different species. For example, honeyguide birds lead badgers to bees' nests.

Some niches are very simple, and the animals that occupy them live simple, solitary lives. Others, especially those occupied by mammals, are much more complex and require members of a species to live closely together. These aggregations of animals may be simple herds or more structured social groups.

Food chain

Food chains show how the energy needed for life passes through an ecosystem. The energy originates in the Sun. This makes plants grow, which are then eaten by animals. The plant-eating animals then become meals themselves.

Below: This food chain shows what animals eat in a temperate region. Herbivores eat only plants, while carnivores eat mainly other animals. Animals that eat both plants and animals are omnivores – for example, humans.

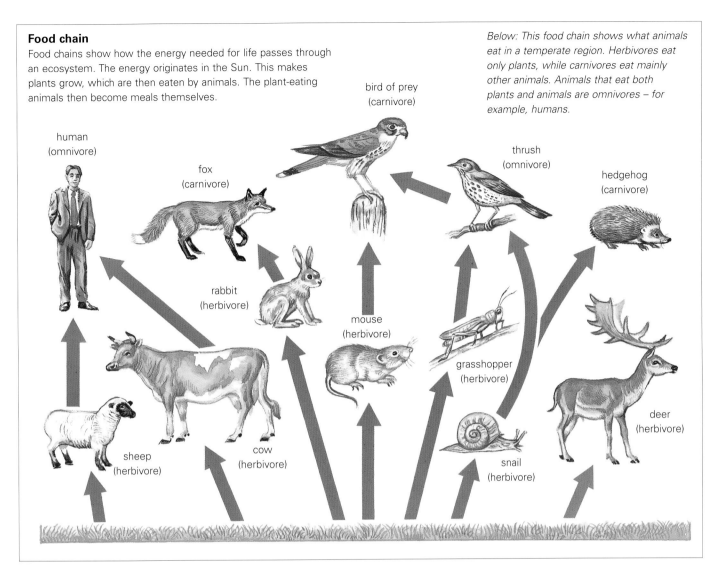

Group living

A herd, flock or shoal is a group of animals which gathers together for safety. Each member operates as an individual, but is physically safest in the centre of the group, the danger of attack being greatest on the edge. Herd members do not actively communicate dangers to each other. When one is startled by something and bolts, the rest will probably follow.

Members of a social group, on the other hand, work together to find food, raise their young and defend themselves. Many mammals, for example apes, monkeys, dogs, dolphins and elephants, form social groups, and these groups exist in many forms. At one end of the spectrum are highly ordered societies, such as lion prides and baboon troops, which are often controlled by one dominant male, the other members often having their own ranking in a strict hierarchical structure. At the other end of the spectrum are leaderless gangs of animals, such as squirrel monkeys, which merge and split with no real guiding purpose.

There are many advantages of living in social groups, with members finding more food and being warned of danger, for example. However, in many societies only a handful of high-ranking members are allowed to breed. In these cases, the groups are held together by a complex fusion of family ties in which brothers and sisters help to raise nephews and nieces. Politics also plays its cohesive part, with members forming and breaking alliances in order to rise to the top.

Below: Plant-eating marsupials such as pademelons often browse together in groups of up to four individuals, but if they are disturbed they tend to scatter as a way of confusing would-be predators.

INTRODUCED SPECIES

Centuries ago, as people started exploring and conquering new lands, many animals travelled with them. In fact, that's the only way many animals could travel such long distances, often crossing seas. Many introduced species then thrived in their new habitats, often at the expense of the native wildlife.

Perhaps one of the first introduced animals was the dingo. This Australian dog is now regarded as a species in its own right, but it was introduced to the region by the Aboriginals. These migrants had domesticated the ancestors of the dingo, probably from grey wolves, many years before.

Dingoes were one of the first placental mammals to come to Australia, which had previously been populated by marsupial mammals. In a chain of events that has been repeated many times since, the introduced dingoes became feral (living as wild animals), and were soon competing with native hunters for food. Not only did the introduced mammals begin to take over from the marsupial predators, such as Tasmanian devils and marsupial wolves, but they also wiped out many smaller marsupials which were unable to defend themselves against these ferocious foreign hunters. In fact, many more of the native Australian marsupials have now become extinct, or are in danger of doing so, though not just dingoes are responsible. As European settlers have arrived over the last couple of centuries, they have also introduced many new animals, including cats.

Domestic animals

Looking around the European countryside, you would be forgiven for thinking that cows, sheep and other farm animals are naturally occurring species. In fact, all come from distant parts of the world. Over the centuries, livestock animals have been selectively bred to develop desirable characteristics, such as lean meat or high milk production. Despite this, they can be traced back to ancestral species. For example, goats – a domestic breed of an Asian ibex – were introduced to North Africa about 3,000 years ago. These goats, with their voracious appetites, did well feeding on dry scrubland. In fact they did too well, and had soon stripped the plants, turning even more of North Africa into desert. Similarly, horses introduced to the Americas by European settlers had

Left: Dingoes are wild dogs that live in Australia. However, they did not evolve there; no large placental mammals did. They are actually the relatives of pet dogs that travelled with the first people to reach Australia.

Cows and sheep

European cows are believed to be descendants of a now-extinct species of oxen called the auroch, while modern sheep are descended from the mouflon. From their beginnings in the Middle East, new breeds were introduced to all corners of the world, where they had a huge effect on the native animals and wildlife.

Above: Cattle have been bred to look and behave very differently from their wild ancestors. Few breeds have horns, and they are generally docile animals. Some breeds produce a good supply of milk, while others are bred for their meat.

Below: Sheep were one of the first domestic animals. They are kept for their meat and sometimes milk, which is used for cheese. The thick coat or fleece that kept their ancestors warm on mountain slopes is now used to make woollen garments.

Above: While the ancestors of domestic horses have become extinct, horses have become wild in several parts of the world. Perhaps the most celebrated of these feral horses are the mustangs, which run free in the wild American West, after escaping from early European settlers.

a marked effect on that continent. Native people, as well as the settlers, began to use them to hunt bison on the plains, which eventually gave way to cattle. Groups of the introduced horses escaped from captivity and began to live wild. The feral American horses were called mustangs.

Rodent invaders

While many animals were introduced to new areas on purpose as livestock or pets, other animals hitched a lift. For example, some animals were more or less stowaways on ships, but only those that could fend for themselves at their new destination were successfully introduced. These species tended to be generalist feeders, and none was more successful than rodents, such as mice and rats. In fact the house mouse is the second most widespread mammal of all,

after humans. It lives almost everywhere that people do, except in the icy polar regions, although it is very likely that rodents did reach these places but then failed to thrive in the cold.

The black rat – also known as the ship rat – has spread right around the world from Asia over the last 2,000 years. On several occasions it has brought diseases with it, including bubonic plague, or the Black Death, which has killed millions of people. Another prolific travelling rodent is the brown rat which is thought to have

spread from Europe, and now exists everywhere except the poles.

Rodents are so successful because they will eat almost anything and can reproduce at a prolific rate. These two characteristics have meant that mice and rats have become pests wherever they breed.

Below: With their sharp and ever-growing teeth, rodents are very adaptable animals. Mice and rats have spread alongside humans, and wherever people go, these little gnawing beasts soon become established, breeding very quickly and spreading into new areas.

CONSERVING WILDLIFE

With so many species facing extinction, conservationists have their work cut out. Conservationists try to protect habitats and provide safe places for threatened animals to thrive, but the activities of ordinary people can often also have an adverse effect on the future of natural habitats.

People give many reasons why wildlife should be conserved. Some argue that if all the forests were cleared and the oceans polluted, the delicate balance of nature would be so ruined that Earth would not be able to support any life, including humans. Others suggest that if vulnerable species were allowed to die, the natural world would not be sufficiently diverse to cope with future changes in the environment. Another reason to save diversity is that we have not yet fully recorded it. Also, there are undoubtedly many as yet unknown species – especially of plants – which could be useful to humankind, for example in the field of medicine. But perhaps the strongest argument for the conservation of wildlife is that it would be totally irresponsible to let it disappear.

Habitat protection

Whatever the reasons, the best way to protect species in danger of being wiped out is to protect their habitats so that the complex communities of

plants and animals can continue to live. However, with the human population growing so rapidly, people are often forced to choose between promoting their own interests and protecting wildlife. Of course, people invariably put themselves first, which means that the conservationists have to employ a range of techniques in order to save wildlife.

In many countries it has now become illegal to hunt certain endangered animals, or to trade in any products made from their bodies. Whales, gorillas and elephants are protected in this way. Many governments and charitable organizations have also set up wildlife reserves, where the animals stand a good chance of thriving. The oldest

Below: One of the main causes of deforestation is people clearing the trees and burning them to make way for farmland. The ash makes good soil for a few years, but eventually the nutrients needed by the crops run out and so the farmers often begin to clear more forest.

Above: If logging is done properly, it can make enough money to pay to protect the rest of the rainforest. Only selected trees are cut down and they are removed without damaging younger growth. Forests can be used to grow crops, such as coffee and nuts, without cutting down all the trees.

protected areas are in North America and Europe, where it is illegal to ruin areas of forest wilderness and wetland. Consequently, these places have become wildlife havens. Other protected areas include semi-natural landscapes which double as beauty spots and tourist attractions. Although these areas often have to be extensively altered and managed to meet the needs of the visitors, most still support wildlife communities.

In the developing world, wildlife refuges are a newer phenomenon. Huge areas of Africa's savannahs are protected and populated with many amazing animals. However, the enormous size of these parks makes it very hard to protect the animals, especially elephants and rhinoceroses, from poachers.

Reintroduction

Large areas of tropical forests are now protected in countries such as Brazil and Costa Rica, but often conservation efforts come too late because many animals have either become rare or are completely absent after years of human

Zoo animals

Once zoos were places where exotic animals were merely put on display. Such establishments were increasingly regarded as cruel. Today, the world's best zoos are an integral part of conservation. Several animals, which are classified as extinct in the wild, can only be found in zoos where they are being bred. These breeding programmes are heavily controlled to make sure that closely related animals do not breed with each other. Later, individual animals may be sent around the world to mate in different zoos to avoid in-breeding.

Below: Many of the world's rarest species, such as the red panda, are kept in zoos, partly so that people can see them, since they are too rare to be spotted in the wild. Some people are opposed to animals being put on display for this reason.

damage. However, several conservation programmes have reintroduced animals bred in zoos into the wild.

To reintroduce a group of zoo-bred animals successfully into the wild, conservationists need to know how the animal fits into the habitat and interacts with the other animals living there. In addition, for example when trying to reintroduce orang-utans to the forests of Borneo, people have to teach the young animals how to find food and fend for themselves.

Below: Breeding centres are an important way of increasing the number of rare animals. Most, such as this giant panda centre in China, are in the natural habitat. If the animals kept there are treated properly, they should be able to fend for themselves when released back into the wild.

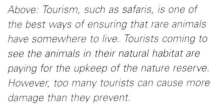

Above: Tourism, such as safaris, is one of the best ways of ensuring that rare animals have somewhere to live. Tourists coming to see the animals in their natural habitat are paying for the upkeep of the nature reserve. However, too many tourists can cause more damage than they prevent.

Understanding habitats

A full understanding of how animals live in the wild is also vitally important when conservationists are working in a habitat that has been damaged by human activity. For example, in areas of rainforest which are being heavily logged, the trees are often divided into isolated islands of growth surrounded by cleared ground. This altered habitat is no good for monkeys, which need large areas of forest to swing through throughout the year. The solution is to plant strips of forest to connect the islands of untouched habitat, creating a continuous mass again.

Another example of beneficial human intervention involves protecting rare frogs in the process of migrating to a breeding pond. If their migration necessitates crossing a busy road, it is likely that many of them will be run over. Conservationists now dig little tunnels under the roads so that the frogs can travel in safety. Similar protection schemes have been set up for hedgehogs and ducks, to allow them safe passageways.

CLASSIFICATION

Scientists classify all living things into categories. Members of each category share features with each other – traits that set them apart from other animals. Over the years, a tree of categories and subcategories has been pieced together, showing how all living things seem to be related to each other.

Taxonomy, the scientific discipline of categorizing organisms, aims to classify and order the millions of animals on Earth so that we can better understand them and their relationship to each other. The Greek philosopher Aristotle was among the first people to do this for animals in the 4th century BC. In the 18th century, Swedish naturalist Carolus Linnaeus formulated the system that we use today.

By the end of the 17th century, naturalists had noticed that many animals seemed to have several close relatives that resembled one another. For example lions, lynxes and domestic cats all seemed more similar to each other than they did to dogs or horses. However, all of these animals shared common features that they did not share with frogs, slugs or wasps.

Linnaeus devised a way of classifying these observations. The system he set up – known as the Linnaean system – orders animals in a hierarchy of divisions. From the largest division to the smallest, this system is as follows: kingdom, phylum, class, order, family, genus, species.

Each species is given a two-word scientific name, derived from Latin and Greek. For example, *Panthera leo* is the scientific name of the lion. The first word is the genus name, while the second is the species name. Therefore *Panthera leo* means the *"leo"* species in the genus *"Panthera"*. This system of two-word classification is known as binomial nomenclature.

Below: The tiger (Panthera tigris) *belongs to the genus* Panthera, *the big cats, to which lions also belong. All cats are members of the larger order of* Carnivora *(carnivores) within the class of* Mammalia *(mammals).*

Above: Like the tiger, the leopard (Panthera pardus) *belongs to the cat family,* Felidae. *This is divided into two main groups, the big and small cats. The group of small cats includes jungle cats and domestic cats.*

Lions, lynxes and other genera of cats belong to the *Felidae* family. The *Felidae* are included in the order *Carnivora*, along with dogs and other similar predators. The *Carnivora*, in turn, belong to the class *Mammalia*, which also includes horses and all other mammals.

Mammals belong to the phylum *Chordata*, the major group which contains all vertebrates, including reptiles, amphibians, birds, fish and some other small animals called tunicates and lancelets. In their turn, *Chordata* belong to the kingdom *Animalia*, comprising around 31 living phyla, including *Mollusca*, which contains the slugs, and *Arthropoda*, which contains wasps and other insects.

Although we still use Linnaean grouping, modern taxonomy is worked out in very different ways from the ones Linnaeus used. Linnaeus and others after him classified animals by their outward appearance. Although they were generally correct when it came

Close relations

Cheetahs, caracals and ocelots all belong to the cat family *Felidae*, which also includes lions, tigers, wildcats, lynxes and jaguars. Although big cats can generally be distinguished from *Felis* by their size, there are exceptions. For example the cheetah is often classed as a big cat, but it is actually smaller than the cougar, a small cat. One of the main behavioural differences between the two groups is that big cats can roar but not purr continuously, while small cats are able to purr but not roar.

Above: The cheetah (Acinonyx jubatus) *differs from all other cats in possessing retractable claws without sheaths. This species is classed in a group of its own, but is often included within the group of big cats.*

Above: The caracal (Caracal caracal) *is included in the group of small cats, (subfamily* Felinae), *but most scientists place it in a genus of its own,* Caracal, *rather than in the main genus,* Felis.

Above: The ocelot (Felis pardalis) *is a medium-sized member of the* Felis *or small cat genus. Like many cats, this species has evolved a spotted coat to provide camouflage – unfortunately attractive to hunters.*

Distant relations

All vertebrates (backboned animals) including birds, reptiles and mammals such as seals and dolphins, are thought to have evolved from common fish ancestors that swam in the oceans some 400 million years ago. Later, one group of fish developed limb-like organs and came on to the land, where they slowly evolved into amphibians and later reptiles, which in turn gave rise to mammals. Later, seals and dolphins returned to the oceans and their limbs evolved into paddle-like flippers.

Above: Fish are an ancient group of aquatic animals that mainly propel themselves by thrashing their vertically aligned caudal fin, or tail, and steer using their fins.

Above: In seals, the four limbs have evolved into flippers that make highly effective paddles in water but are less useful on land, where seals are ungainly in their movements.

Above: Whales and dolphins never come on land, and their ancestors' hind limbs have all but disappeared. They resemble fish but the tail is horizontally – not vertically – aligned.

to the large divisions, this method was not foolproof. For example, some early scientists believed that whales and dolphins, with their fins and streamlined bodies, were types of fish and not mammals at all. Today, accurate classification of the various genera is achieved through a field of study called cladistics. This uses genetic analysis to check how animals are related by evolutionary change. So animals are grouped according to how they evolved, with each division sharing a common ancestor somewhere in the past. As the classification of living organims improves, so does our understanding of the evolution of life on Earth and our place within this process.

DIRECTORY OF ANIMALS

Asia is the world's largest continent, stretching halfway around the globe, while Australia, an island between the Indian Ocean and the Pacific, is the smallest continent. Being an island, Australia's wildlife has been isolated for many millions of years and today the animals that live there are very different from those found in Asia and beyond. While Asia is home to mammals such as elephants, orang-utans, wild horses and camels, Australia is populated by kangaroos, wombats and possums. Most of Australia's mammals carry their young in pouches, and a few even lay eggs like reptiles. The reptiles and amphibians of Australia are more similar to those found in Asia, but they are by no means ordinary. Perhaps the most unusual reptile of all, the tuatara, lives in New Zealand, a group of islands in the South Pacific. These ancient reptiles have no living relatives anywhere else in the world.

Above from left: Carpet python, pretty face wallaby, koala.

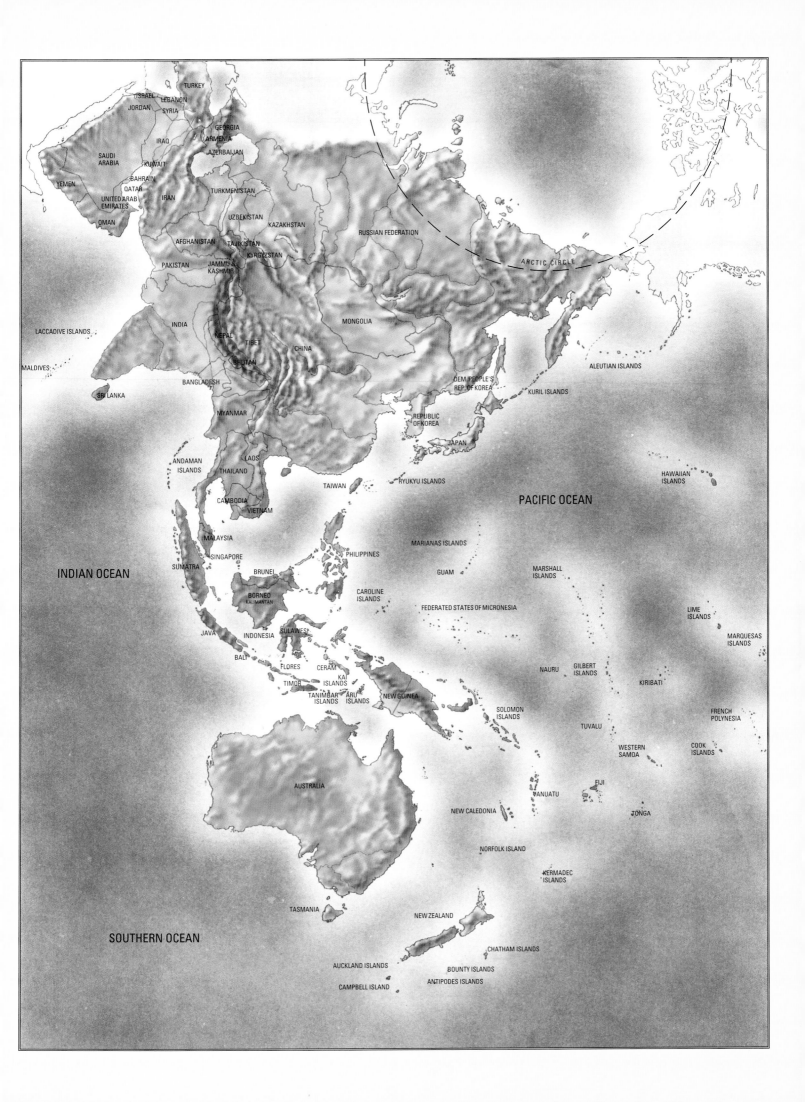

SALAMANDERS AND RELATIVES

Salamanders can be found in both aquatic and terrestrial habitats, although they need water for survival during the early stages of their lives. Young salamanders breathe through external gills. Once they reach adulthood, salamanders develop lungs and start to breathe air, but they can also absorb oxygen through their skins, which allows them to spend long periods underwater.

Ceylon caecilian

Ichthyophis glutinosus

Distribution: Sri Lanka.
Habitat: Underground in damp forest soils.
Food: Insect larvae, termites and earthworms.
Size: 30–50cm (12–20in).
Maturity: Not known.
Breeding: 7–20 live young born after gestation of 9–11 months.
Life span: 6–7 years.
Status: Common.

Caecilians look like snakes or worms, but they are neither. In fact they are amphibians, being closely related to frogs, toads and salamanders. Unlike most amphibians, which have few or no teeth, caecilians have two sets of backward-pointing teeth on their upper and lower jaws.

While being relatively common in their forest habitats, Ceylon caecilians are difficult animals to find. This is because they spend virtually their entire lives under the thick leaf litter of tropical forests, away from the sharp eyes of the many birds, mammals and reptiles that would eat them.

Being blind, caecilians find food and mates using smell. They have small tentacle-like appendages that protrude from just behind their nostrils. These tentacles smell the environment immediately around them, detecting their prey – insects and earthworms – and mates. Females produce strong odours when they are fertile.

With eyes covered by skin and bone, caecilians are totally blind. Their slender body shape makes them perfectly adapted for a subterranean, burrowing lifestyle. The family holds over 90 species in all.

Japanese giant salamander

Andrias japonicus

This salamander is one of Japan's most impressive inhabitants, and one of the world's largest salamanders. It belongs to a small group of salamanders called *Cryptobranchidae*, to which American hellbenders also belong. Weighing over 20kg (44lb) and reaching sizes well in excess of 1m (3.25ft), the Japan giant salamander is found only in cold, fast-flowing mountain streams. It spends the daytime hiding under rocks and branches, coming out at night to feed on its prey, which it catches using a sticky tongue and powerful jaws.

This creature actually has three names in Japanese; they are "Osanshouuo", "Hanzaki", and "Hajikamiio". These names are derived from the fact that the salamanders secrete mucus from their skins that smells like the Japanese peppers called "sansho" and "hajikami". Until recently the Japanese giant salamander was a highly prized source of food for people living in the mountainous regions of Japan. However, the pressures of over-hunting and water pollution have caused the population to drop very sharply, and now the giant salamander is strictly protected by law.

Distribution: Japan.
Habitat: Cold, fast-flowing mountain streams with rocky bottoms.
Food: Mainly worms and insects, and also crabs, fish, frogs, snakes and mice.
Size: 40–70cm (16–28in), up to 140cm (56in); over 20kg (44lb).
Maturity: 5–10 years.
Breeding: 500–600 eggs laid in autumn, hatching after 40–50 days.
Life span: 70 years.
Status: Vulnerable.

This huge salamander is dark brown with black spots. It has characteristically warty skin on its head. Its pepper-like smell is caused by mucus from glands in the skin.

Western Chinese mountain salamander

Batrachuperus pinchonii

This sturdy salamander inhabits the fast-flowing streams of the mountainous Sichuan province of China. These streams are too cold for a lot of animals, especially fish, which tend to be a major predator of salamander larvae. Therefore, the young salamanders grow up in relatively predator-free environments; they just have to contend with the cold.

This species' tail is relatively short for a salamander, and is flattened from side to side – laterally – into a paddle. This helps the salamander swim against the fast currents. Eggs are laid under stones so that they do not get washed away. In the very cold conditions, the salamanders have low metabolic rates and do not move very quickly, appearing a bit slow and ponderous. They take many years to reach maturity in their chilly habitat.

The Western Chinese mountain salamander is a robust amphibian with a mainly glossy, brown or green skin, generally with spots. Its diet consists mainly of insects.

Distribution: Sichuan Province, China.
Habitat: Fast-flowing mountain streams.
Food: Mainly insects.
Size: 15–20cm (6–8in).
Maturity: Not known.
Breeding: 7–12 eggs laid per year.
Life span: Not known.
Status: Locally common.

Paddletail newt (*Pachytriton labiatus*): 12–15cm (4.8–6in)
Found only in southern China, the paddletail newt is aptly named. It has a very broad, flat tail, which helps it to swim rapidly through the slow-flowing rivers in which it lives. It has very smooth, shiny skin which is dark brown along its back, but bright red with black patterning on its belly. It is a very secretive animal, spending most of its time hiding away under rocks and in crevices, only coming out at night to feed.

Afghan salamander (*Batrachuperus mustersi*): 12–17cm (4.8–6.8in)
Found only in the Paghman Mountains of Afghanistan, this salamander inhabits cool highland streams fed by glaciers. Indeed, it seems that it cannot survive in streams that are warmer than 14°C (57°F). The Afghan salamander is entirely aquatic and has a dark olive-brown coloration, with little in the way of markings. The restricted range of this animal means that it is highly endangered, and it is feared that it may have been badly affected by the recent conflicts in Afghanistan.

Chinese fire belly newt (*Cynops orientalis*): 10cm (4in)
Found throughout most of China, fire belly newts are striking little animals. Although they are dark grey or brown on top, when they are turned over it is easy to see how they got their name. Their bellies are coloured bright red, with small black spots. Chinese fire belly newts spend most of their time in water, but come on to land occasionally in search of the small invertebrates that make up their diet. In the breeding season, females lay single eggs on the leaves of water weeds and then fold the leaves over to hide the eggs.

Crocodile newt

Tylototriton verrucosus

The crocodile newt is named after the raised spots that line its back, reminiscent of the raised scutes of crocodiles. However, the newt's bumps are bright orange as a warning to predators that they secrete foul-tasting toxins. Crocodile newts leave the water at the onset of winter, when they bury themselves into soft soils to hibernate. During their active period – in spring, summer and autumn – the newts only come on land when water has become too scarce.

Like most newts, crocodile newts have a very acute sense of smell, being able to find food even in total darkness. They are mainly nocturnal and often forage in muddy pools. These newts can locate prey by smell alone when the density of mud makes their sense of sight useless. They also use smell to attract their mates. The males release a sex-smell which attracts the females to them. Males then perform courtship dances to induce the females into picking up their packages of sperm, known as spermatophores, which they deposit as they dance in circular patterns.

The crocodile newt has black skin, an orange head and limbs, and raised orange spots down its spine and back.

Distribution: Mountains of south-east and southern Asia.
Habitat: Aquatic habitats, mainly shallow temporary pools.
Food: Worms and insects.
Size: 12–22cm (4.8–8.8in).
Maturity: 1 year.
Breeding: 30–150 eggs laid, which take about 2 weeks to hatch.
Life span: 12 years.
Status: Endangered.

FROGS AND TOADS

Frogs and toads, collectively known as anurans, are by far the best known of the three groups of amphibian. Adult anurans are characterized by their lack of a tail, which is lost in the metamorphosis from tadpole to adult. Tadpoles tend to feed mainly on algae and plant matter, whereas adults tend to eat insects and other animals. Adult frogs are also characterized by their strong rear legs, used for jumping.

Oriental fire-bellied toad

Bombina orientalis

Oriental fire-bellied toads have green backs with black blotches. Their bellies are orange-red with black mottling.

Despite the remarkably vivid colouring of its underbelly, the fire-bellied toad merges perfectly with its surroundings; its dull brown to bright green back renders it almost invisible against the muddy margins of the pools and ponds where it lives. The back is also usually dotted with glossy black spots, which further help to camouflage it. Its skin gives off a milky secretion which irritates the mouth and eyes of would-be pedators.

When the fire-bellied toad feels threatened, it has a surprising trick, known as the unkenreflex. The toad flips over on to its back and arches its body, revealing the bright warning colours on its belly. These colours warn potential predators that the toad can secrete poisonous chemicals from its skin. When a predator has tasted these secretions once it will rarely try and attack a fire-bellied toad again. However, some snakes and birds do appear to be able to eat the toads with no ill-effects.

Distribution: Korea.
Habitat: Warm, humid forest areas, usually near water.
Food: Invertebrates, including worms, molluscs and insects.
Size: 4–5cm (1.6–2in).
Maturity: 2 years.
Breeding: 40–250 eggs laid in batches of 3–50 eggs throughout the summer. Hatch in June–July.
Life span: 30 years.
Status: Common.

Asian horned frog

Megophrys montana

Distribution: Thailand, Malaysia, Indonesia and the Philippines.
Habitat: Rainforest.
Food: Insects.
Size: 10–13cm (4–5.2in).
Maturity: Not known.
Breeding: Not known.
Life span: Unknown.
Status: Common, but declining due to habitat loss.

The Asian horned frog inhabits the dense rainforests of South-east Asia. It is a master of disguise, and might easily be mistaken for a leaf. A long snout, large eyelids and skin the colour of dried leaves all help the animal to stay hidden on the forest floor. By doing this it can remain motionless, in relative safety from predators, while waiting for its next meal to walk past.

The tadpole of the Asian horned frog has evolved a curious specialization that concerns its mouth. The tadpole gets most of its food from the surface of the pool or stream that it is living in. Most tadpoles have downward-pointing mouths, but this would make it difficult for this particular species to reach food, so it has evolved a mouth that points upwards instead. The tadpole is therefore able to cruise just beneath the surface of the water and pick off the tiny plants on the surface.

The points of skin above the eyes constitute the "horns" of the Asian horned frog. These and its leaf-like coloration help it to camouflage itself among fallen leaves on the forest floor.

Southern gastric brooding frog

Rheobatrachus silus

There are two species of gastric brooding frog, both now feared to be extinct in the wild. They are the only species in the world to share a particularly bizarre method of looking after their young. Immediately after laying her eggs, the adult female frog, somewhat surprisingly, swallows them. However, she has the extraordinary ability to switch off the production of digestive juices, and she turns her belly into a nursery for six or seven weeks, during which time she is unable to eat. The eggs hatch, and the tadpoles develop to near-maturity within the safety of their mother's stomach, before being "born again" through her mouth.

Southern gastric brooding frogs were only discovered in 1973, at which time they were thought to be relatively common. However, during the following few decades a combination of habitat destruction, increased pollution levels and over-collection for the pet trade have seen this incredible species driven into extinction in the wild. The species now only survives in captivity.

This small, dull-coloured frog with large bulging eyes has a special method for looking after its young. They develop within the mother's stomach before emerging out of her mouth.

Distribution: Queensland, Australia.
Habitat: Rocky creek beds and rainforest rock pools.
Food: Small insects.
Size: 3–5.5cm (1.2–2.2in).
Maturity: 1 year.
Breeding: 18–30 eggs are brooded for 6–7 weeks within the stomach of the adult female.
Life span: 3 years.
Status: Extinct in the wild, but survives in captivity.

Stonemason toadlet (*Uperoleia lithomoda*): 1.5–3cm (0.6–1.2in)
This tiny frog is found only in the northernmost areas of Australia. Emerging from its underground burrow only when the infrequent rains arrive, the stonemason toadlet has to find a mate rapidly and produce young before conditions again become too dry. The male produces a strange click-like courtship call which sounds like two stones being knocked together. It is this call which gave the stonemason toadlet its name.

Asian painted frog (*Kaloula pulchra*): 5–7cm (2–2.8in)
After a heavy bout of rain, the deep, booming bellows of this frog can be heard from most drains, culverts, lakes and ponds throughout much of South-east Asia. This chubby little frog lives up to its name, being coloured with browns, blacks and beige. When threatened, it puffs itself up with air, and contorts its face as a warning to any would-be predator.

Brown-striped frog (*Limnodynastes peronii*): 3–6.5cm (1.2–2.5in)
Also known as the Australian grass frog, this long-limbed amphibian with dark stripes running down its body lives in and around swamps in eastern Australia. This amphibian is nocturnal in its habits, feeding at night on insects and other invertebrates. During the winter and dry periods, it buries itself underground. During mating, pairs make floating foam nests into which 700–1,000 eggs are laid. The eggs hatch quickly and the tadpoles develop very rapidly, too.

Giant tree frog

Litoria infrafrenata

Found in forests and gardens, the giant tree frog is Australia's largest native species of frog, and one of the largest tree frogs in the world. As its name suggests, this frog spends most of its life in the branches of tall trees, feeding on insects at night. During the day, when it is more visible to predators such as birds of prey, it stays well out of sight in foliage. Its green coloration helps to camouflage it as it hides away.

Following spring storms, males gather in trees around swamps, lakes and ponds and produce bizarre calls in an attempt to attract females to mate. The mating calls resemble the deep-throated barks of large dogs, but it isn't just dogs that the frogs can sound like. When distressed, giant tree frogs may make loud meowing sounds, just like the distress calls of cats.

Once the eggs have been laid they develop quickly, hatching within 28 hours. Hatching so quickly means that the young only spend a short period of time lying motionless and prone to predation within their eggs.

Usually an olive or emerald-green shade, the giant tree frog has the ability to change colour depending on temperature and the surrounding vegetation.

Distribution: Northern Australia to Indonesia.
Habitat: Occurs in a variety of habitats, ranging from tropical forest to cultivated areas.
Food: Insects.
Size: 14cm (5.6in).
Maturity: Not known.
Breeding: Between 200 and 300 eggs laid in spring and summer.
Life span: Not known.
Status: Common.

Cane toad

Bufo marinus

Distribution: Originally from the Amazon Basin, now introduced into Australia, the United States, the West Indies, Puerto Rico, Taiwan, Hawaii, the Philippines and many other islands.
Habitat: Wide-ranging and adaptable to many different habitats.
Food: Insects and small vertebrates.
Size: 24 cm (10in); 1.8kg (4lb).
Maturity: 1 year.
Breeding: 2 clutches of between 8,000 and 35,000 eggs produced each year. Eggs hatch into free-swimming tadpoles that become adult in 45–55 days.
Life span: 40 years.
Status: Common.

The cane toad, or marine toad, is now most famous for being a serious pest. Originally native to the Amazon Basin, it was deliberately released into Australia in the 1930s in the hope that it would control the grey-backed beetle, a pest of sugar cane. Unfortunately the cane toad preyed on almost everything except the grey-backed beetle, which stayed out of reach high up the stems of cane crops. Now a number of Australian species, including other frogs and toads, are in danger of dying out after being extensively preyed upon by these toads.

The cane toad's success in both its native habitat and in new areas where it has spread is helped by the fact that it has two glands just above its shoulder blades, which produce large quantities of highly toxic venom. This venom is so potent that snakes, which have tried to eat the toads, have been found dead with the toads still lodged in their mouths.

Snakes are not the only animals affected. Native marsupial predators such as quolls, along with introduced cats and dogs, are all at risk if they try to eat the toads. Cane toads also push native tree frogs out of the best breeding grounds, further establishing their reputation as a serious problem for Australia's native wildlife.

The large, brown cane toad has become a common sight in eastern Australia in recent years. The amphibian has a large, wide head with bony ridges on the top and sides, large protruding eyes, and a stout, powerful body. The glands on its shoulder blades produce a venom that is capable of killing goannas, crocodiles, snakes, dingoes, quolls, cats, dogs and humans within the space of 15 minutes. Cane toads eat almost anything, including pet food, carrion and household scraps, as well as their main diet of insects and small vertebrates.

Holy cross frog

Notaden bennetti

The holy cross frog is also known as the catholic frog or crucifix frog. A native of Australia, it is a yellow or greenish amphibian, and its name is derived from the dark cross-like pattern on its back. It is a desert-dwelling frog, surviving the heat by burrowing underground.

Summer in the Australian desert is not a hospitable season. The sun bakes the ground with great ferocity and there is virtually no water to be found anywhere. However, despite these harsh conditions, the holy cross frog is able to survive. It is active for only a few weeks every year, just after the rains, when it emerges from its burrow and busily sets about the business of finding a mate, reproducing and preparing to sit out another summer.

When it rains, holy cross frogs are able to absorb water through glands in their skins. They can absorb as much as half of their body weight, and become as round as balls. Then, when the rains stop, they burrow 30cm (12in) or more under the desert floor, where they stay until the next rainfall. During this time they virtually shut down completely, with all of their vital life processes running at the lowest possible level to make sure that they do not use any more water than they need to. This survival mechanism is called aestivation.

Distribution: Australia.
Habitat: Widespread throughout arid regions.
Food: Small black ants form the main element of the diet.
Size: 4cm (1.6in).
Maturity: 1 year.
Breeding: Occurs after heavy rains in summer.
Life span: Not known.
Status: Common.

Hochstetter's frog

Leiopelma hochstetteri

The Hochstetter's frog is believed to be one of the world's most primitive frogs. It is one of the few species of frog that shares some specific physical characteristics with fish, which are relatives of the amphibians. For example, it has no external ear openings, and no vocal sac. Also, the frog does not croak, and can only manage a quiet squeak at the best of times. Additionally, it has two tail-wagging muscles, but no tail.

Hochstetter's frog lives mainly under rocks and boulders near mountain streams and other wet places in New Zealand. In fact, it is one of only three species of frog native to this region. It is still relatively common in some localized areas of the North Island, whereas the other two species are exceedingly rare, confined to just a few offshore islands.

The tadpoles of this species are unusual in that they don't feed until they change into adult frogs. When a female lays her eggs, she does so into a large, water-filled capsule. After hatching, the tadpoles stay within this capsule until they mature.

Distribution: North Island, New Zealand.
Habitat: Under stones or vegetation alongside creeks.
Food: Insects, spiders, worms and slugs.
Size: 5cm (2in).
Maturity: 1 year.
Breeding: Egg clusters laid in damp, shady places.
Life span: Not known.
Status: Threatened.

This small brown frog is believed to be one of the most primitive of all frog species.

Corroboree frog

Pseudophryne corroboree

Distribution: Kosciuszko National Park in New South Wales, Australia.
Habitat: Bogs and surrounding woodland.
Food: Insects.
Size: 2.5–3cm (1–1.2in).
Maturity: 2 years.
Breeding: 16–38 eggs laid in summer, hatching 7 months later.
Life span: Not known.
Status: Critically endangered.

The strikingly beautiful corroboree frog is thought to be Australia's – and possibly one of the world's – most endangered species of frog. Less than 200 individuals are still alive, isolated from each other in small groups numbering no more than 25. It is believed that corroboree populations have been decimated by a disease known as chytrid (pronounced kit-rid).

The frogs are particularly at risk due to their slightly peculiar lifestyle. Corroboree frogs have much slower breeding cycles than other frogs. While tadpoles are fully formed just four weeks after the female has laid her eggs, they don't actually hatch out of the eggs for another six months. Having finally broken out of their eggs, the tadpoles then take a long time to reach maturity and become adult frogs.

This slow lifestyle means that it takes corroboree frog numbers longer to recover from a population crash than it would other frogs. The population is at a very low ebb today, and is in great danger of dying out entirely.

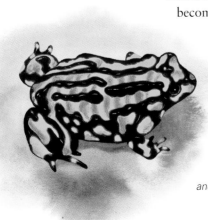

The corroboree frog is a spectacularly colourful species, with well-defined black and yellow markings.

Solomon Islands horned frog (*Ceratobatrachus guentheri*): 5–8cm (2–3.2in)
Unusually for a frog, this species has tooth-like projections growing from its lower jaw. It uses these to keep hold of its prey, such as worms and insects. The projections may also play some role in competition between males for mates, because they are larger in the males than they are in females. After mating, the females lay batches of eggs which develop directly into miniature adult frogs, with the tadpole stage occurring within the eggs.

Fijian ground frog (*Platymantis vitiensis*): 3.5–5cm (1.4–2in)
Confined to the forests of Fiji, little is known about the habits of this small orange-brown frog. It has very prominent eyes and large eardrums, as well as large pads on its toes, which help it to climb around. The female lays a small number of large eggs directly on to the ground. The young develop directly into young adults, the tadpole stage being completed within the eggs. Fijian ground frogs are highly endangered due to their limited original habitats being deforested.

Ornate narrow-mouthed frog (*Microhyla ornata*): 2–3cm (0.8–1.2in)
The ornate narrow-mouthed frog inhabits the rainforests and paddy fields of South-east Asia, where it feeds on insects and other invertebrates. This small frog has a plump, smooth-skinned body, with short front legs, and long back legs. It is mainly coloured yellow, with dark stripes along the back and a darker marbled pattern on its back.

Asian blue-web flying frog

Rhacophorus reinwardtii

Flying frogs don't fly as such – they glide downwards from tree to tree using the webbing between their toes. When feeling threatened, these frogs will leap from their current branch into the air. They spread their toes, expanding the flaps of blue skin between them, and use these as parachutes. This technique allows the frogs to glide more than 12m (40ft) between trees, giving them a useful technique for escaping predators.

This species of flying frog has an interesting breeding method. The females create foam nests in which they lay their eggs. These nests are positioned above pools of water that collect in leaves high up in the forest canopy. The foam prevents the eggs from drying out while the tadpoles are developing within. Once sufficiently mature, the tadpoles then break out of their foam nests and drop into the water below, where they complete their development into adult frogs. The generally green coloration provides camouflage in the forest canopy.

Well camouflaged by their green skins amongst the leaves, the blue webbing between the flying frog's toes provides a startling contrast when they jump.

Distribution: Indonesia, Malaysia and South-east Asia.
Habitat: Rainforest.
Food: Insects.
Size: 4–7cm (1.6–2.8in).
Maturity: Not known.
Breeding: Clutches of up to 800 eggs in a foam nest placed above a pool of water.
Life span: 5 years.
Status: Common.

Indonesian floating frog (*Occidozyga lima*): 10–12cm (4–4.8in)
The Indonesian floating frog spends most of its time simply bobbing about in the ponds and lakes in which it lives. It holds on to a piece of weed with one or more feet, looking like a fallen leaf resting on the surface of the water. When insect prey come close enough, they are gobbled up with quick lunges.

Black-spined toad (*Bufo melanostictus*): 10cm (4in)
One of Asia's most successful amphibians, the range of the black-spined toad stretches from India in the west to Hong Kong in the east and Indonesia in the south. It has been able to colonize cultivated and urban areas, making use of drainage ditches, ponds and paddy fields. These toads have numerous warts on their backs, and it is these that have given them their name. Black-spined toads produce venomous secretions from their warts, which also have the effect of staining their skin black.

Crab-eating frog (*Ratia cancrivora*): 8–10cm (3.2–4in)
The crab-eating frog is one of the few species of amphibian that can tolerate living in salty, brackish water. Inhabiting brackish waters brings the frog into contact with prey that frogs would not usually encounter, crabs being a good example. As their name suggests, crab-eating frogs will often feast on crabs when they get the chance, but they are not fussy, and will eat pretty much any prey that is small enough for them to swallow.

White's tree frog

Litoria caerulea

It may look as though the blue-green skin of White's tree frog is ill-fitting, but it is this skin which is the frog's key to success. During very wet conditions, the amphibian is able to absorb large amounts of water and store it under the loose folds of skin. The water can then be used in times of drought. However, water-storing is not the only talent possessed by this frog's skin. It also produces various anti-bacterial compounds, and a chemical which has been shown to be useful for treating high blood pressure in humans.

During the mating season, males often cling to the backs of females for days on end, waiting for them to lay eggs, which are then laid in an unusual fashion. Instead of carefully placing their eggs into water, like many frogs, female White's tree frogs squirt their eggs out underwater with such great force that they may travel for more than 2m (6ft) before coming to a stop. White's tree frogs are able to produce some very loud sounds. When alarmed, the frogs will emit ear-piercingly loud screams – enough to put off many predators.

Distribution: Australia and southern New Guinea.
Habitat: Moist, forested environments are preferred, but can live in drier areas.
Food: Moths, locusts, cockroaches and other insects.
Size: 7–11.5cm (2.8–4.6in).
Maturity: 2 years.
Breeding: 1,500–3,000 eggs hatch within 2 or 3 days of being laid.
Life span: 21 years.
Status: Common.

The White's tree frog has a waxy, blue-green colour and rolling skin folds of fatty material that have earned it the nickname of "dumpy tree frog".

Giant frog

Cyclorana australis

Distribution: Northern Australia.
Habitat: Grassland and open woodland. Can survive in more arid areas than most species.
Food: Insects.
Size: 7–11cm (2.8–4.4in).
Maturity: 2 years.
Breeding: Occurs after the summer rainy season. 7,000 eggs laid in temporary ponds. Tadpoles mature quickly before pools dry up.
Life span: Not known.
Status: Common.

The giant frog is a species of frog that is able to live in much drier areas than most frogs. It survives the long, dry days of an Australian winter by cocooning itself in a mixture of shed skin, mucus and damp earth. It can also retain large amounts of water beneath its skin and in its bladder and body cavity – an adaptation which helps it to sit out long dry spells in the safety of its cocoon.

The ability to hold large amounts of water has made the giant frog, and its relative the water-holding frog, a valuable animal for Aboriginal people, who dig them up and squeeze the pure water out of the frogs to gain a drink. The frogs are released lacking water supplies but otherwise unharmed.

Once the summer rains arrive, the frogs leap into action, feasting on the large numbers of insects and preparing to mate. The males gather around the edges of temporary pools and call to females, which come down and lay up to 7,000 eggs. The young have to hatch and develop into adult frogs within the space of a few short weeks, before their pools of water dry up in the hot sun.

The giant frog's large eyes give it excellent vision, which it uses to spot insects such as beetles, grasshoppers and termites. It is a nocturnal, burrowing frog, and is usually only seen above ground after rain. These amphibians of dry terrain in Australia are much prized by the local Aboriginal people, who use them to gain moisture when water is scarce.

Asian tree toad

Pedostibes hosii

The Asian tree toad, also known as the brown tree toad, is a large and stocky toad that lives in lowland tropical forest below 660m (2,165ft) above sea level. It is distributed throughout Borneo and is also known in Sumatra, Peninsular Malaysia and southern Thailand. The tadpoles develop in quiet pools of water and at the bottoms of streams, amongst dead leaves. When they become adults, the toads develop arboreal skills and spend some of their time in trees, hence their name.

This toad is unusual among toads in its extraordinary ability to climb. While most toads lead purely terrestrial lives, the Asian tree toad has sucker-like pads on its toes, which enable it to climb with ease. It spends its days hidden away in crevices in trees and in foliage, only becoming active when the sun sets.

Throughout the night, the tree toad comes down to the forest floor along rivers and streams and feeds mainly on ants, which it picks up with its long, sticky tongue. The animal calls to other toads by making a grating, slurred squawk that rises slightly in pitch from beginning to end.

The female lays her eggs in strings in rivers. These rivers need to be fairly fast-flowing because the eggs need to be kept in conditions where the concentration of oxygen in the water is relatively high. Once they hatch, however, the tadpoles have to battle against the fast water currents to prevent themselves from being washed away. They have large sucker-like mouths with which they cling to rocks, anchoring themselves against the flow.

Distribution: South-east Asia.
Habitat: Lowland tropical forest.
Food: Insects, mainly ants.
Size: 5–11cm (2–4.4in).
Maturity: 2 years.
Breeding: Many small eggs laid in a string in rivers.
Life span: 16 years.
Status: Common.

The Asian tree toad varies in colour from greenish-brown to black, with a dense pattern of yellow spots. Some females are dark purple. Unlike most toads, this South-east Asian species is a skilled climber, shinning up trees with the aid of the sticky pads on its toes.

TURTLES

Among the oldest surviving groups of reptiles, turtles first evolved about 200 million years ago. They have changed very little since then, a fact that bears testament to their successful body design and lifestyle. Turtles live both on land and in water, with the land-living species being more commonly referred to as tortoises. All species of turtle lay their eggs on land.

Common snake-necked turtle

Chelodina longicollis

The neck of the common snake-necked turtle is very long. When extended, the neck is longer than half the length of the shell.

With their long, flexible neck, good eyesight and strong jaws, snake-necked turtles are formidable predators in their aquatic habitats. They are cosmopolitan animals, feeding on any creatures they can catch, including frogs, tadpoles and fish. Breeding occurs in September and October – the spring months in Australia – and in November the females lay clutches of eggs in holes in the banks of swamps or ponds. They then cover the holes over and leave the eggs to develop on their own. After three to five months the young turtles hatch and dig themselves out of the holes, scurrying down to the comparative safety of the water.

Most snake-necked turtles live close to streams, rivers, swamps and lagoons in eastern Australia. From time to time, they migrate over land in groups to search out new habitats. It is not unusual for these turtles to colonize artificial ponds.

Distribution: South-eastern and eastern Australia.
Habitat: Swamps, lakes, slow-moving waterways, creeks and billabongs.
Food: Frogs, tadpoles, small fish and crustaceans.
Size: 30cm (12in); 1.2kg (2.6lb).
Maturity: 4–5 years.
Breeding: Clutches of 8–24 eggs hatch after 3–5 months.
Life span: 50 years.
Status: Common.

Chinese soft-shelled turtle

Pelodiscus sinensis

Distribution: China, Korea and Japan.
Habitat: Rivers, lakes, ponds and reservoirs.
Food: Snails, molluscs, crabs, fish, shrimp, insects, frogs and earthworms.
Size: 20–35cm (8–14in).
Maturity: Not known.
Breeding: Eggs hatch 28 days after laying.
Life span: Not known.
Status: Endangered.

The Chinese soft-shelled turtle inhabits slow-flowing rivers and ponds with sandy or muddy bottoms. Occasionally it can be found out of water, basking on stones, but it is quick to disappear underwater when it feels threatened. The Chinese soft-shelled turtle is a predominantly nocturnal animal, foraging on the riverbed at night for prey. This turtle is extremely cosmopolitan in its tastes, eating most animals that it can get its powerful jaws around. Crayfish, snails, insects, fish and amphibians are all on the menu during a night's foraging.

The soft-shelled turtle's jaws are also useful for defence. Unlike most types of turtles, this species cannot fully retract its head under its shell. It also lacks the hardened, bony plates on the shell which protect other turtles. Instead it has a vicious bite, enough to protect it from most would-be predators. The turtle's long snout acts as a snorkel, allowing it to stay completely submerged in shallow water and yet still breathe.

The Chinese soft-shelled turtle has a flat, soft and rubbery shell and a long snout which acts as a snorkel.

Indian starred tortoise (*Geochelone elegans*):
24–27cm (9.6–10.8in)
One of the world's most distinctive species of
tortoise, the Indian starred tortoise is found only
on the Indian subcontinent. It has a bizarrely
shaped and beautifully marked shell. Each
section of the shell, known as a scute, rises to a
domed point. These tortoises are mainly active
in the monsoon season, when there is plenty of
moisture around. During the driest season, they
are active only in the mornings and evenings.

Pig-nosed river turtle (*Carettochelys insculpta*):
70–75cm (28–30in)
Found only in the far north of Australia, the
pig-nosed river turtle is unlike most freshwater
turtles in that it has broad flipper-like limbs, more
reminiscent of sea-living turtles. Its name refers
to its pig-like snout, which is used for breathing
while the rest of the body remains underwater.

Chinese box turtle (*Cuora flavomarginata*):
10–12cm (4–4.8in)
The Chinese box turtle has a hinged line
along its plastron (the underside of its shell).
This allows it to close up entirely when it feels
in danger. Found only in a small area of eastern
Asia, this endangered species spends most
of its time in paddy fields feeding on fish,
crustaceans, worms and fruit.

Asian leaf turtle

Cyclemys dentata

The early part of an Asian leaf turtle's life is
spent almost entirely in water. The young
are virtually exclusively aquatic. However,
as they grow older the turtles spend
increasing amounts of time on land. This
species of turtle is omnivorous in habits,
feeding on both plant and animal matter. It
will quite happily switch from eating leaves
to picking off passing snails or crickets.

The lower half of a turtle's shell is called
the plastron, and in most turtles this
features a series of radiating lines. In the
Asian leaf turtle the plastron is hinged
along the length of the animal in such
a way that, when threatened, the
turtle can pull its head and legs fully
into its shell, and then pull the
lower part of the shell up to close
off the holes. This forms a near-
impenetrable barrier, behind which
the turtle can take cover.

*The Asian leaf turtle has a flatter shell than
most turtles. It is usually dark brown in colour,
but can be lighter.*

Distribution: South-east Asia.
Habitat: Streams.
Food: Invertebrates, tadpoles
and some plant material.
Size: 15–24cm (6–9.6in).
Maturity: Not known.
Breeding: 5 clutches of
2–4 eggs laid each year.
Life span: Not known.
Status: Not known.

TUATARA

*While looking like a typical lizard, the tuatara is actually a member of a completely separate group
of reptiles known as the* Rhynchocephalia, *of which it is the sole remaining example.*

Tuatara

Sphenodon punctatus

*Tuataras are often described as
living fossils. They are the only
surviving members of a group of
reptiles that otherwise died out
long ago.*

The *Rhynchocephalia* flourished about 200 million years
ago, living alongside the dinosaurs. However, most of the
species suffered the same fate as the dinosaurs: extinction.
The only members to survive were the tuataras. There are
only two species still living, both found in New Zealand.

Unlike lizards, tuataras have a third eye. This is a
light-sensitive organ lying just under the skin on the
top of the head. When they are young, the third eye
absorbs ultraviolet rays, helping the youngsters
produce vitamin D – essential for bone growth.
Nearly everything about the lifestyles of tuataras is
slow. Eggs develop within the female's body for
about ten months before they are laid. It then takes
another year for the baby tuataras to hatch. They
do not become sexually mature until the age of 20
years. Tuataras have very slow metabolic rates, and
breathe on average only once every seven seconds.
However, they can hold their breath for over an hour.

Distribution: Coastal islands
of New Zealand.
Habitat: Underground
burrows.
Food: Insects, small lizards
and birds' eggs and chicks.
Size: 50–60cm (20–24in);
0.5kg (1.1lb).
Maturity: 20 years.
Breeding: 6–10 eggs once
every 2–5 years.
Life span: 100 years.
Status: Lower risk.

LIZARDS

The most diverse and widespread group of reptiles is made up of the 3,000 species of lizards. They vary in size from just a few centimetres to almost 3m (10ft) long, and exhibit huge variation in shape, coloration and feeding habits. From legless burrowers to forest dwellers that "fly" from tree to tree, lizards manage to carve themselves a niche almost everywhere on earth.

Frilled lizard

Chlamydosaurus kingii

Distribution: Northern Australia and southern New Guinea.
Habitat: Woodland.
Food: Predominantly insects, but also small lizards and mammals.
Size: 55–70cm (22–28in).
Maturity: 2–3 years.
Breeding: 2 clutches of 10–20 eggs produced every year.
Life span: 20 years.
Status: Common.

Possibly one of the most spectacular lizards in the world, this animal uses its incredible frill to frighten would-be predators into believing that it is dangerous.

In normal circumstances, the frilled lizard looks somewhat similar to other large lizards. It is drab-coloured, being able to blend into its background and remain hidden from its predators. However, should this hiding tactic not work, then the lizard's amazing, brightly coloured frill is brought into action. When the lizard opens its mouth, rods of cartilage attached to the jaw cause the frill to open out like an umbrella. At the same time the lizard makes a loud hissing noise. The combination of the loud hiss and opening frill is enough to frighten most predators into believing that this mostly harmless lizard is a fearsome creature not to be messed with.

Frilled lizards forage mainly on the ground but are also very good climbers, being able to chase their insect prey up trees. When on the ground, the frilled lizard has an enviable turn of speed, being able to outrun most humans.

Thorny devil

Moloch horridus

The incredibly spiny appearance of the thorny devil helps to protect and camouflage it, as well as collect water in its harsh desert habitat.

Australia plays host to many strange animals, and the thorny devil most definitely ranks among the most bizarre. Its entire body is covered by a barrage of spines, which are buff or grey in colour. These spines help to break up the lizard's outline, so it is difficult for birds of prey to spot it from above. However, the main benefit of the spines is less obvious – they are a source of water. Thorny devils live in extremely hot and arid areas, where water is at a premium. During the night, condensation collects on the lizard's skin; the spines help to channel this water down into grooves that run along the body to its mouth. This method of collecting water means that the thorny devil never needs to drink directly from a puddle or pond.

Thorny devils eat only ants, which they pick off one by one using their sticky tongues. They are very efficient at this process, being able to pick off up to 50 ants per minute. A single meal may contain as many as 3,000 ants.

Distribution: Western and central Australia.
Habitat: Desert.
Food: Ants.
Size: 15–18cm (6–7.2in); 90g (0.2lb).
Maturity: 3 years.
Breeding: 4–10 eggs laid in summer.
Life span: 20 years.
Status: Lower risk.

Water dragon

Physignathus cocincinus

With a prominent crest running from the back of the head down the spine, the water dragon is a large, impressive lizard. During the breeding season, males are bright green in colour, but become duller at other times of the year.

The water dragon is equally at home in water, on land and even in the trees. However, it will always head very swiftly for the comparative safety of water when it feels threatened.

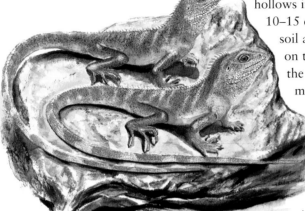

Having mated, females scrape out shallow hollows in which to lay their clutches of 10–15 eggs. They then cover the eggs with soil and leaves, and leave them to develop on their own. After two to three months the young hatch out of the eggs, measuring a mere 15cm (6in). However, they grow quickly and become sexually mature at just one year of age.

The bright green coloration of the male water dragon only lasts during the breeding season, at other times of the year he is duller, resembling the female.

Distribution: South-east Asia: Thailand, Vietnam and Cambodia.
Habitat: Tropical rainforest, near water.
Food: Invertebrates, small mammals, birds, lizards, frogs and fruit.
Size: 80–100cm (32–40in).
Maturity: 1 year.
Breeding: 5 clutches of 10–15 eggs laid each year, hatching after 2–3 months.
Life span: 15 years.
Status: Common.

Sailfin lizard (*Hydrosaurus pustulatus*): 80–100cm (32–40in)
Hailing from the Philippines, sailfin lizards are grey-green in colour. Adults have crests running down their backs and larger crests running down their tails. The tail crests can be up to 8cm (3.2in) high, and help the lizards to swim. Whilst they spend most of their time in trees, they run to water if threatened. The sail-like fin means that they can swim at high speed.

Crocodile lizard (*Shinisaurus crocodilurus*): 40cm (16in)
The only member of its family, the crocodile lizard is found solely in the Guilin and Guangxi regions of China. It gets its name from rows of large olive-green bony scales on its back and tail, which are reminiscent of a crocodile's skin. Local people call this species "the lizard of great sleepiness" because it can remain motionless, in a metabolic pause for hours or days on end.

Bearded lizard (*Pogona vitticeps*): 30–45cm (12–18in)
Inhabiting the dry, arid interior of Australia, bearded lizards are great survivors. They forage for their diet of small insects and vegetation in dry forest and scrubland. These lizards are territorial, defending their home patches from intruders. Within their territories they tend to have regular areas, where they rest and bask in the sun during the early mornings. When confronted with potential predators, bearded lizards puff out their black throat patches and open their mouths wide in the hope of intimidating their foe.

Flying lizard

Draco volans

Also known as flying dragons, flying lizards have sets of wing-like membranes, which are actually skin stretched over elongated ribs. The lizards do not really fly, but glide through the forest from tree to tree. A single glide can be up to 10m (33ft) long. When not gliding, the "wings" are folded away down the sides of the body. Flying lizards have a third flap of skin under their chins, which can be extended at will. These chin flaps are larger and coloured yellow in the males, being used in courtship dances.

These small lizards spend virtually their entire lives in trees. The females only come down to the ground for short amounts of time to bury their eggs in soil. Having laid their eggs and covered over the holes, the females will fiercely defend the eggs for approximately 24 hours, before returning to the trees and leaving the young to develop independently. Having hatched, the young immediately climb the nearest trees to start their arboreal lives; the males will probably never set foot on the ground again.

Distribution: South-east Asia, Indonesia, the Philippines and Borneo.
Habitat: Forest and woodland.
Food: Ants and other small insects.
Size: 15–22cm (6–8.8in).
Maturity: 1 year.
Breeding: Clutches of 5 eggs laid in holes in the ground.
Life span: 8 years.
Status: Common.

The skin stretched over its elongated ribs gives the flying lizard a pair of wing-like membranes, which it uses to glide from tree to tree.

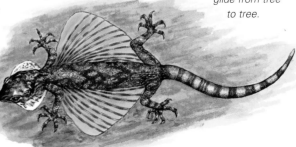

Tokay gecko

Gekko gecko

The tokay gecko is among the largest species of gecko in the world, reaching lengths of almost 40cm (16in). It gets its name from its most common call, an explosive noise sounding like "to-kay, to-kay, to-kay". The gecko is nocturnal, coming out at dusk to forage for the insects and small vertebrates that make up its diet.

Tokays are fearsome fighters, with powerful bites which are enough to cause considerable discomfort to any predators attempting to make a meal of them. Male tokays are highly territorial and will attack any other males wandering into the wrong area after failing to heed the warnings given by the residents. Tokay fights are fast, furious and generally deadly to their enemies.

Like the majority of gecko species, tokay geckos have specialized pads on the ends of their toes that allow them to grip smooth surfaces with ease. In houses where they are welcomed as a sign of good luck, they climb up windows and across ceilings in search of food. Sometimes they can be found near lights, hoping to catch the moths that are attracted to them at night.

Distribution: South-east Asia.
Habitat: Bushes, trees and on or around rocks. Also commonly found in or near houses, where it is tolerated because it is believed to bring good luck.
Food: Insects, hatchling birds, and also small mammals and reptiles.
Size: 20–37cm (8–15in).
Maturity: 1–2 years.
Breeding: Females lay 1–2 eggs, usually in crevices in rocks or walls.
Life span: 10 years.
Status: Common.

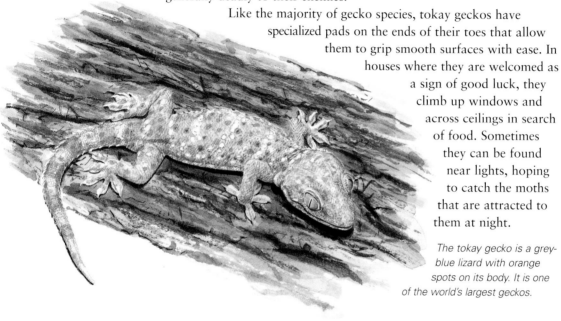

The tokay gecko is a grey-blue lizard with orange spots on its body. It is one of the world's largest geckos.

Flying gecko

Ptychozoon kuhli

Distribution: South-east Asia.
Habitat: Tropical forests.
Food: Insects.
Size: 18–20cm (7–8in).
Maturity: 6 months.
Breeding: 5–6 clutches of 2 or 3 eggs laid throughout the summer.
Life span: 3 years.
Status: Common.

Unlike the flying lizard, the flying gecko uses webbing between its toes as parachutes to glide from tree to tree.

The flying gecko is another species of reptile, along with the flying snake and flying lizard, that has taken to gliding around the forests of South-east Asia by using flaps of skin as airfoils. However, unlike the other "flying" reptiles, the flying gecko doesn't use flaps of skin stretched over elongated ribs, but instead it has flaps which grow between its long toes, in the same manner as flying frogs. These flaps of skin give the gecko a very fat-handed appearance, and this has given rise to its alternative name, the thick-fingered gecko.

Flying geckos are exceptionally well camouflaged. They closely resemble the bark patterns on the trees upon which they rest motionless, with their heads facing downward for hours on end, waiting for their insect prey to come close enough to grab. They even have lighter patches on their skins, which look like the patches of lichen growing on the tree trunks.

Blue-tongued skink

Tiliqua scincoides

One of Australia's most familiar reptiles, the cosmopolitan habits of the blue-tongued skink mean that it is been able to make the best of most situations. It has a varied diet, including insects, snails, fruit and flowers, as well as carrion and the contents of rubbish bins.

As its name suggests, the blue-tongued skink does indeed have a blue tongue. This is very large, almost the same size as the lizard's head, and when the lizard feels threatened, it will stick its bright tongue out at the attacker, presenting the predator with a warning signal. Combined with a fierce hiss, this display is often enough to discourage all but the most persistent predators.

During the breeding season, males set up territories from which they exclude other males. When females enter their territories, the males chase them around briefly, and then pin them down by the backs of their necks until they have finished mating.

The blue-tongued skink is a stocky reptile with a wide head, thick-set body and short limbs. The blue tongue discourages predators.

Distribution: Eastern Australia.
Habitat: Wide-ranging, including forest and parkland.
Food: Very varied, including insects, molluscs, carrion, fruit, flowers and food scraps from human settlements.
Size: 45–50cm (18–20in).
Maturity: 3 years.
Breeding: Up to 25 young born live.
Life span: 25 years.
Status: Common.

Knob-headed giant gecko (*Rhacodactylus auriculatus*): 20cm (8in)
On the Pacific islands of New Caledonia there is a group of bizarre geckos that have evolved in isolation for many thousands of years. The knob-headed giant gecko is one of them, and is so named because it has a series of raised crests on its skull that give it a knobbed or horned appearance. It is highly variable in colour, from being mottled to having striking broad bands of brown and white running down its body. Unlike most geckos, which are predominantly insectivorous, adult knob-headed giant geckos eat mainly fruit.

Leach's giant gecko (*Rhacodactylus leachianus*): 30cm (12in)
This gecko is among the world's largest and is reputedly the only gecko in the world that growls. This strange nocturnal call may be one of the reasons why some locals consider this and other giant geckos to be "devils in the trees".

Long-tailed skink (*Mabuya longicauda*): 30–35cm (12–14in)
Living in open habitats such as gardens and parks, often near water, the long-tailed skink is aptly named. Its tail may be twice as long as the rest of its body and head combined. The long-tailed skink lives in South-east Asia, where it is a common sight rooting around leaf litter in search of insects and spiders. Female long-tailed skinks do not lay eggs, but give birth to live young. The young are miniatures of their parents, and receive no parental care after they are born.

Emerald tree skink

Lamprolepis smaragdina

Few reptiles can claim to be as brightly coloured as the emerald tree skink, which inhabits the tropical forests of South-east Asia and New Guinea. It is an adept climber, spending the majority of its time hiding among the dense foliage on tree trunks and branches, rarely venturing down to the ground. The emerald tree skink actively forages for its insect prey, which it catches with rapid, darting movements.

Like a lot of lizards, this species has the ability to shed its tail if grabbed – a phenomenon known as autotomy. This is a very useful tactic when attacked by predators. Once shed, the tail carries on moving and will often distract the predator, usually a bird of prey or snake, long enough for the rest of the lizard to make its escape.

Distribution: South-east Asia and New Guinea.
Habitat: Tropical forest.
Food: Insects.
Size: 18–25cm (7.2–10in).
Maturity: 6 months.
Breeding: Clutches of 9–14 eggs laid in summer.
Life span: 4 years.
Status: Common.

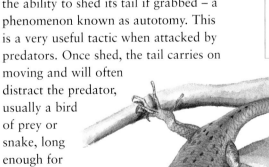

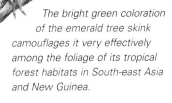

The bright green coloration of the emerald tree skink camouflages it very effectively among the foliage of its tropical forest habitats in South-east Asia and New Guinea.

Sand monitor

Varanus flavirufus

The sand monitor, or goanna, is one of the largest Australian lizards. It is also sometimes known as the racehorse monitor because it is capable of extreme speeds, often running only on its two back legs, using its long tail for balance.

Sand monitors are found throughout most of Australia, and have managed to be very successful by turning most situations into opportunities. They will eat virtually anything, searching for morsels of food with their long forked tongues, which are used for tasting the air. They will even dig up the nests of crocodiles when the mother's back is turned, and feast on the eggs within.

These lizards are not overtly aggressive, preferring to stay out of trouble whenever possible, but they can put up a serious fight if need be. They use their long whip-like tails, powerful jaws and long slashing claws to defend themselves from predators. Their claws also allow them to climb trees, usually in search of birds' nests.

Distribution: Australia.
Habitat: Woodland, shrubland and grassland.
Food: Lizards, insects, spiders, scorpions, centipedes and even small mammals and carrion.
Size: Up to 1.6m (5.5ft).
Maturity: Not known.
Breeding: 10 eggs laid into a deep burrow.
Life span: Not known.
Status: Common.

Sand monitors have flattened bodies, long powerful tails, stout limbs, long fingers and very sharp claws. Their well-developed hind legs help them achieve high speeds.

Lace monitor (*Varanus varius*): 1.5–2m (5–6.5ft)
Found in eastern Australia, the lace monitor is dark grey in appearance, with whitish markings, a long neck and tail and long, sharp claws, which it uses to climb trees. The long tail helps the monitor to balance while it is climbing, and it also can become a useful weapon when faced with a predator.

Borneo earless lizard
(*Lanthanotus borneensis*): 40–45cm (16–18in)
This lizard is nocturnal and highly secretive, and therefore its elongated red-brown form is rarely seen in its native range of north-west Borneo. The earless lizard belongs to a family of its own, but is most closely related to the monitor lizard. It is a good swimmer and a proficient burrower, having a long muscular tail, short stocky legs and a blunt snout. It is able to close its nostrils when diving underwater. Other than this, little is known about the lifestyle of this intriguing species.

Green tree monitor

Varanus prasinus

The green tree monitor blends wonderfully well into the foliage of the trees where it spends most of its days. It also has other adaptations that help it live an arboreal life, such as thick, rough scales on the soles of its feet, which help it to grip branches. Most importantly, it has a highly dextrous tail, which acts like a fifth limb while climbing. Many species of monitor will use their tails as a whip during defence. However, the tail is such an important organ to the green tree monitor that it actively defends it from predators.

Not much is known about the lifestyle of this lizard, due to its secretive nature and the inaccessibility of its habitat. However, it is known that it forages in the forest canopy, mainly for large insects. Some of these insects, such as some katydids, are very well protected with long spiny legs. Before attempting to eat these prickly morsels, green tree monitors therefore tend to strip the legs off and discard them, eating only the nutritious and less spiky body parts.

Distribution: New Guinea.
Habitat: Rainforests, palm forests, mangroves and cocoa plantations.
Food: Large insects and the occasional small mammal.
Size: Up to 1m (3.25ft).
Maturity: 2 years.
Breeding: 2 or 3 clutches of up to 5 eggs produced per year.
Life span: Not known.
Status: Common.

A long, graceful lizard with a pointed head and a whip-like tail, the green tree monitor can be many shades of green, from pale to very dark, depending on where it lives.

Komodo dragon

Varanus komodoensis

The Komodo dragon is probably the world's most infamous lizard. Not only is it the largest and most aggressive living lizard, but it is also one of the most endangered. It belongs to the same family as the monitor lizards, but has evolved separately for millions of years, becoming incredibly large in the process.

The Komodo dragon is just one of many giant species of animals that have evolved on islands around the world. It is also thought that Komodo dragons originally became so big that they were able to prey on a species of pygmy elephant that inhabited Indonesia thousands of years ago. These pygmy elephants have since become extinct, and the dragons now prey on other mammals, all of which have been introduced into their islands by human populations. The dragons lie hidden for hours upon end, waiting for prey animals to wander past, and then spring from their ambush positions with incredible bursts of speed.

Usually the prey is overpowered within seconds. However, should it escape, the problems for the victim are not yet over. Komodo dragons have a wealth of poisonous bacteria living in their mouths. When they bite prey animals, these bacteria are introduced to the wound, and often cause it to go septic and fester. This results in a slow, painful death.

Female Komodo dragons lay an average of about 20–25 soft, leathery eggs in September. About twice the size of chicken's eggs, they incubate for 8–9 months during the wet season. The young hatch out and immediately start to look for insect prey. However, they have to be careful, because they could well end up on the menus of larger dragons.

Distribution: Indonesian islands of Komodo, Rintja, Gillimontang, Padar and the western tip of Flores.
Habitat: Dry savannah and woodland.
Food: As adults, the bulk of the diet is made up of large mammals: goats, deer, pigs, horses and water buffalo. Birds and reptiles will also be taken.
Size: 3m (10ft); 150kg (330lb).
Maturity: 5 years.
Breeding: 20–25 eggs laid, incubation 8–9 months.
Life span: 30 years.
Status: Endangered.

This is the world's largest lizard. The huge form of the Komodo dragon belies the speed with which it can move.

Burton's snake lizard

Lialis burtonis

Distribution: Australia and southern New Guinea.
Habitat: Wide-ranging, from forests to desert.
Food: Lizards, mainly skinks.
Size: Up to 60cm (24in).
Maturity: Not known.
Breeding: 1 clutch of 2 eggs laid per year.
Life span: Not known.
Status: Common.

While it may look like a snake, this species is actually a legless lizard. Furthermore, it is a lizard that feeds on other lizards, which it catches in a cunning way. Having located a likely victim, the Burton's snake lizard keeps its body very still, but raises the tip of its tail slightly, and wiggles it. The skinks, upon which it feeds, are curious animals and are lured over to investigate. While the skink is preoccupied with the wiggling tail, it fails to notice the head watching it intently from a few centimetres away. Once in range, the snake lizard strikes, holding its prey tightly in its mouth until it suffocates to death. Burton's snake lizard likes to be hot; if the outside temperature drops down to about 21°C (70°F), it will enter an inactive (dormant) state in order to save energy.

Very variable in colour and markings, Burton's snake lizard is legless and has a broad, blunt snout used for burrowing.

CROCODILIANS

Little-changed since the time of the dinosaurs, crocodiles are notorious for their stealth and ferocity while on the hunt. Despite their bloodthirsty reputation, female crocodiles are caring mothers. They guard their nest sites, help the young hatch from their eggs and carry them to the water. Even though they spend much of their time in water, crocodiles are also effective on the land, and can outrun humans.

Saltwater crocodile

Crocodylus porosus

The saltwater crocodile is probably the most fearsome reptile in the world. Better education is encouraging people to stay away from these giant animals.

This dark grey reptile is the largest of the crocodile species and the heaviest reptile in the world. A full-grown adult can weigh over 1,000kg (2,200lb) and be as much as 7m (23ft) long. It has a large head and powerful jaws, designed for holding and crushing. The crocodile eats pebbles, which are thought to aid digestion by grinding food, and they also act as a ballast for maintaining buoyancy.

The female "saltie" builds a nest from earth and grasses, which keeps the eggs safe from flooding. She guards the eggs from predators and carries the new hatchlings to the water.

Juvenile crocodiles eat crustaceans, fish and reptiles, but as they grow larger, they often take creatures as big as a buffalo or domestic livestock. Humans are occasionally killed or injured.

Distribution: South-east Asia and Northern Australia.
Habitat: River mouths and swamps.
Food: Mammals, birds and fish.
Size: 5–7m (16–23ft); 500–1,000kg (1,100–2,200lb).
Maturity: 10–12 years.
Breeding: 40–60 eggs laid in summer.
Life span: 70 years.
Status: Common.

Gharial

Gavialis gangeticus

The gharial is easily recognized by its long, slender snout, which is filled with interlocking razor-sharp teeth. The male has a bulbous growth on the tip of the nose, used for making vocalizations and producing air bubbles when underwater. It uses bubble displays to attract females.

The gharial is poorly adapted to life on land because its leg muscles are not suitable for walking, and therefore it spends most of its life in water. It prefers quiet river backwaters, where its flattened tail and webbed hind feet make swimming easy.

Fish are the most common food for the gharial. The narrow jaws are well designed for quick snapping motions underwater, and the victims are swallowed head-first. Juveniles will often also eat small crustaceans and frogs. Gharials almost became extinct during the 1970s, and today they are protected throughout much of their range.

Distribution: Northern India, Bangladesh and Pakistan.
Habitat: Wide, calm rivers.
Food: Fish.
Size: 4–7m (13–23ft).
Maturity: 10 years.
Breeding: 30–50 eggs laid from March–May.
Life span: Not known.
Status: Endangered.

The bizarre-looking slender snout of the gharial helps it to grip its slippery fish prey. The reptile rarely comes on land except to nest.

Mugger

Crocodylus palustris

As its name suggests, the mugger crocodile rarely passes up an opportunity to grab a meal. Adults capture prey as large as buffalo.

This medium-sized crocodile is named from a Hindu word meaning "water-monster", and true to its name, the reptile poses quite a threat to any prey that catches its eye. The young crocodiles will take mostly small crustaceans and fish, but the adults will also prey on snakes, turtles and even deer and buffalo. They prefer slow-moving fresh water, not more than around 5m (16ft) deep, and will travel long distances over land in search of water during the dry season.

Mugger crocodiles look very similar to alligators, with broad snouts and flattened heads. However, their sharp teeth are perfectly aligned, and this is what distinguishes them from the alligators in appearance. They have flattened tails and are strong swimmers, but do not use their feet to swim, despite their being webbed.

Like all "crocs", muggers are social animals. The adults and juveniles often call to each other and, during the mating season, males establish territories and dominance by raising their snouts high above the ground and thrashing their tails.

Distribution: India, Pakistan and Sri Lanka.
Habitat: Shallow, slow-moving fresh water.
Food: Juveniles feed on fish, frogs and crustaceans; adults take a wide range of prey, including reptiles and mammals such as deer and buffalo.
Size: 2–5m (6.5–16ft).
Maturity: Females 6 years; males 10 years.
Breeding: Up to 28 eggs laid between February and April.
Life span: 40 years.
Status: Threatened.

Chinese alligator

Alligator sinensis

Distribution: Yangtze River in China.
Habitat: Slow-moving fresh water.
Food: Aquatic invertebrates and fish.
Size: 2m (6.5ft); 40kg (88lb).
Maturity: 4–5 years.
Breeding: 10–50 eggs laid from July–August.
Life span: Not known.
Status: Critically endangered.

Young Chinese alligators have bold markings of yellow and black, but these fade as the alligators mature, becoming more and more grey as time passes.

This small and secretive alligator is found only in one small part of China, along the banks of the Yangtze River. There are believed to be less than 200 Chinese alligators alive in the wild. Protection is in place, but to many farmers these alligators are nothing more than a nuisance, prone to attacking their wildfowl. Consequently, many are killed on sight. The loss of suitable habitat is also a major problem, and many captive breeding programmes have been postponed because there is not the space for the new specimens to be reintroduced to the wild.

Chinese alligators grow to around 2m (6.5ft) long and weigh up to 40kg (88lb). They are usually dark green or black in colour. Juveniles have bright yellow cross-banding. In common with the more familiar American species, *Alligator mississippiensis*, Chinese alligators have long, upturned snouts and bony ridges along their bodies and tails.

The teeth of Chinese alligators are well adapted for crushing, as they feed extensively on crustaceans and molluscs. They hunt mostly at night and during the summer months. They are opportunistic feeders, not averse to catching ducks or rats if they find them. They will rarely attack larger mammals.

SNAKES

The highly evolved snakes make fearsome and deadly predators of animals of all sizes, from insects to antelope. Snakes cannot chew their food, but can disassociate the two halves of their jaws and swallow their prey whole. Some species of snake produce the most potent natural toxins in the world, however they are in the minority, and only a small fraction of these reptiles pose any threat to humans.

Banded sea krait

Laticauda colubrina

Distribution: Pacific and Indian Oceans from Japan to Australia and Africa.
Habitat: Shallow coastal waters.
Food: Eels and fish.
Size: Typically 1–1.5m (3.3–5ft), up to 3.6m (12ft).
Maturity: 18–24 months.
Breeding: Clutches of 8–20 eggs laid on land.
Life span: Not known.
Status: Common.

The banded sea krait is a tropical marine snake, well adapted to life underwater and on land. It is a venomous but docile snake with distinctive black and yellow bands along its body and a creamy underbelly. It spends much of the time in water, but returns to the land to breed and shed its skin. On land, well-developed stomach scales known as scutes give the snake considerable terrestrial agility and tree-climbing ability, but once egg-laying is over, it returns to the sea.

Sea kraits stay mainly in shallow, tropical waters around South-east Asia and Australia, where they eat eels and small fish. They have flattened paddle-shaped tails for swimming, nostril flaps to keep out the water and specialized glands under their tongues for excreting salt from sea water. Average males, at 0.7cm (2.5ft) long, are dwarfed by females, which can easily reach 1.25m (4ft). As a result, females often venture into deeper waters in search of larger eels, avoiding competition for food.

Silver and black stripes cover the entire body of the banded sea krait, from the neck area to the tip of the tail.

Taipan

Oxyuranus scutellatus

The taipan of Australia is one of the most venomous snakes in the world, with a bite that can kill a mouse in three seconds and an adult human in 30 minutes. It is highly aggressive and employs a snap-and-release attack strategy, whereby a thrust and sudden bite is followed by withdrawal to avoid being crushed by the possible death throes of the victim.

Taipans thrive on the northern coasts of Australia, possibly because they only eat mammals. When the highly toxic cane toad was introduced to the area, it caused the numbers of frog-eating snakes to decline, leaving plenty of prey and vacant habitat for the taipan. Its lethal bite and athletic lunging have earned it the nickname the "poison pogo-stick".

The female taipan becomes sexually mature at a smaller size than the male, and will often lay two clutches of 10–20 eggs during the breeding season. The hatchlings are around 45cm (18in) long. They are usually a light olive colour, with creamy underbellies and reddish eyes.

Distribution: North Australian coast.
Habitat: Open grassland and woodland.
Food: Small mammals.
Size: 1.5m (5ft).
Maturity: 6–12 months.
Breeding: Females lay 1 or 2 clutches of 10–20 eggs per year.
Life span: Not known.
Status: Common.

Taipans are very variable in colour, ranging from sandy and light russet through to jet black.

Malayan pit viper

Calloselasma rhodostoma

The Malayan pit viper probably bites more humans each year than any other species of snake in Asia, and there is a good reason why this is so. It is a nocturnal hunter and spends the day curled up in leaf litter on the forest floor, well camouflaged by its red-brown triangular markings. Unfortunately this behavioural trait means that the snake will not move out of the way or give any kind of warning if a human is walking towards it; it will bite.

Like all vipers, the Malayan pit viper has long, hinged fangs, which lie pointing backwards when the snake has its mouth closed, and snap forward into position when the snake opens its mouth and rears up to strike. It has heat-sensing pits located between its eyes and nostrils, which are so sensitive they can detect the warmth given off by the bodies of their prey. This allows the snake to hunt in darkness.

A well-camouflaged and deadly beauty, the Malayan pit viper is one of Asia's most dangerous snakes.

Distribution: South-east Asia.
Habitat: Rubber plantations and rainforests.
Food: Small mammals.
Size: 0.7–1m (2.3–3.3ft).
Maturity: 1 year.
Breeding: 2 clutches of 13–25 eggs laid per year.
Life span: Not known.
Status: Common.

Red-tailed pipe snake (*Cylindrophis rufus*): 0.7–1m (2.3–3.3ft)
This harmless pipe snake is usually found underground but sometimes takes to water, actively hunting small eels and other snakes. When attacked, it raises its tail, showing the red underside, thus mimicking a cobra raising its head. Not only does this make the snake appear more dangerous than it actually is, but it also confuses predators as to which end is the head and which is the tail.

Sunbeam snake (*Xenopeltis unicolor*): 1–1.3m (3.3–4.3ft)
Under dull light this snake appears dark brown to black, but under sunlight it displays an iridescence of red, blue, green and yellow. The highly polished, glimmering scales give the sunbeam snake its name. It spends most of its time underground, coming out at night to feed on frogs, lizards and small mammals. It is found throughout southern China and South-east Asia.

Pacific ground boa (*Candoia carinata*): 0.7–1m (2.3–3.3ft)
Highly variable in colour and patterning, the Pacific ground boa – from Papua New Guinea and the Solomon Islands – can be striped, blotched or banded. However, most individuals have a flowery, blotched pattern and range in colour from yellow, grey and beige to black or even orange and red. The ground boas living in the Solomon Islands never leave the ground, while those found in New Guinea will climb trees and bushes quite readily in search of the small animals upon which they feed.

King cobra

Ophiophagus hannah

The king cobra is one of the largest snakes in the world, sometimes reaching 5.5m (18ft). It can rear up as tall as a human. It has flaps of loose skin around the head that can be flattened out to form a narrow hood. The king cobra is not strictly a true cobra and, as such, has an unmarked hood.

Depending on its habitat, the king cobra may vary its colour, and it is often darker when in forest than when in open savannah. It has good eyesight and will chase down fleeing prey over long distances. Unusually, king cobras eat only reptiles – mostly other snakes.

The intelligent king cobra is the only snake known to build a nest, which is fiercely guarded by the female. Once hatched, the young are just as deadly as the adults, ready to hunt from only ten days old. The smaller young are vulnerable to mongooses and giant centipedes, but the only threat to the adults tends to come from humans, who kill them for use in medicinal drugs.

The open hood of a king cobra is a sure sign of trouble. Being able to lift its head over 1.5m (5ft) from the ground, the snake is a fearsome predator. Its scientific name means "eater of snakes".

Distribution: Northern India to Hong Kong and Indonesia.
Habitat: Tropical rainforest and savannah.
Food: Reptiles, especially snakes.
Size: 4–5.5m (13–18ft).
Maturity: 5 years.
Breeding: 20–40 eggs laid between January and April.
Life span: 20 years.
Status: Threatened.

Green tree python

Morelia viridis

Distribution: Papua and Iran Jaya in New Guinea and the Cape York Peninsula of Australia.
Habitat: Tropical forest.
Food: Birds are the main prey item.
Size: 1.8–2.4m (6–8ft).
Maturity: Over 12 years.
Breeding: Females lay clutches of 6–30 eggs in hollow tree trunks.
Life span: Not known.
Status: Common.

Masters of disguise: green tree pythons are virtually impossible to spot among the foliage when they are coiled motionless in a tree.

This beautiful bright green snake spends almost its entire life in the treetops. Most functions – from eating and drinking to mating and egg laying – are performed off the ground. The green coloration gives good camouflage against the leaves, and the prehensile tail lets the snake cling firmly to branches. It will often wait, coiled around a branch and with its head hanging downwards, ready to pounce on passing birds. Vertical pupils make its eyes very sensitive to movement, and the scales around its mouth have heat-sensitive pits which pick up the body heat of animals to aid prey detection.

This species displays the body shape characteristic of pythons: a thick body with a small head, covered with many small scales. Eggs are usually laid in clutches of anywhere from 6–30, in holes in tree trunks. The green coloration of mature snakes is not present in the young, which tend to be bright yellow or occasionally red. They are hunted by birds of prey.

Reticulated python

Python reticulatus

The reticulated python is easily recognized by its unmarked head and sheer size. It is the longest snake in the world. It is a strict carnivore, and kills most effectively by waiting in trees to ambush unsuspecting victims. It tends not to hunt actively, preferring to conserve its energy. It will usually eat birds and small mammals, as well as deer and pigs. Like all reptiles, it has a low metabolic rate and can go for long periods without eating. A captive reticulated python once refused food for 23 months, and then resumed normal feeding.

The reticulated python has a striking net-like pattern of markings, made up of darker triangular shapes along its sides, and patterns of yellow or cream running down its back. The head is largely unmarked, apart from a thin black stripe running across the top.

Reticulated pythons are strong swimmers, although they prefer to spend most of their time on land, hiding in trees. They are nocturnal, and usually breed during the winter months. Large eggs are laid in hollow trees or burrows, then the female coils around them and shivers to keep the eggs warm. Once the baby snakes are hatched, however, they are left to fend for themselves.

Distribution: South-east Asia.
Habitat: Tropical rainforest.
Food: Mainly mammals.
Size: 6–10m (20–33ft); up to 200kg (440lb).
Maturity: 2–4 years.
Breeding: 25–80 eggs.
Life span: 30 years.
Status: Common.

The beauty of the world's longest snake is matched by its strength and power. The reticulated python gets its name from its distinctive skin markings.

Mangrove snake

Boiga dendrophila

Distribution: South-east Asia.
Habitat: Forest near coasts and rivers.
Food: Lizards, birds and frogs.
Size: 2–2.5m (6.5–8.5ft).
Maturity: 1–2 years.
Breeding: Clutches of 4–15 eggs.
Life span: Not known.
Status: Common.

The brightly coloured mangrove snake is easily recognized – it is mainly black, but with bright yellow bands along the length of its body. These bands do not form complete circles around the body, but are interrupted by a pronounced ridge running down the length of the back. It also has yellow lips and throat.

Mangrove snakes are nocturnal hunters, and usually take small vertebrates such as frogs and lizards. They are mildly venomous, and keep their fangs pointing towards the backs of their mouths. As their name suggests, mangrove snakes tend to live in or around mangrove swamps, choosing trees either in coastal waters or near rivers farther inland. They spend most of their time coiled in tree branches and are often fairly docile during the day. The females lay their eggs in piles of rotting leaf litter or old hollow tree stumps.

The bright yellow markings on its jet-black body make the mangrove snake one of the world's most striking reptiles.

Golden tree snake (*Chrysopelea ornata*): 1–1.2m (3.3–4ft)
Also known as the ornate flying snake, this dramatic snake can spread its ribs to form a concave wing shape, allowing it to glide over long distances. When threatened, it will leap from its perch and glide safely to the ground, leaving its hungry predator at the top of the tree. The golden tree snake has only a very mild poison, preferring to kill its mainly lizard prey by crushing it in its powerful jaws.

Tentacled snake (*Erpeton tentaculatum*): 0.7–1m (2.3–3.3ft)
This sluggish, nocturnal snake has two short soft tentacles on its snout. It lives in water and hunts by ambush – lying still and hidden by weeds and waiting for its prey to swim past. When fish or amphibians come within range, the tentacled snake rapidly strikes with a sideways motion, clinging tightly on to its prey before it has a chance to escape.

Blue coral snake (*Maticora bivirgata*): 1.2–1.4m (4–4.5ft)
The highly venomous nature of the blue coral snake is advertised by the bright orange markings on its head and tail. The rest of its body is black, with two beautiful light blue stripes running down each flank. Despite its venom, this species of snake is relatively docile and rarely bites humans, preferring to use its toxins to overpower the other species of snake upon which it feeds. Whilst being primarily nocturnal, it can occasionally be found lying stretched out on forest trails in the early morning, basking in the sunshine.

Red-tailed racer

Gonyosoma oxycephala

The red-tailed racer has a somewhat misleading name. It has a plain green body, fading into a tail which varies in colour from brown to orange or grey, but it is never truly red. It is an arboreal snake, active during the daylight hours. It spends most of its time foraging and hunting for birds, small mammals and bats, searching with its large eyes for any animal sizeable enough to form a good meal.

Its green coloration gives it excellent camouflage while moving through the tropical rainforests where it lives. It rarely comes down from the trees, and even mating takes place in the branches. Despite its relatively small size, the racer is an aggressive snake, always ready to defend itself. When it is threatened by birds or larger mammals, it will inflate its throat to look bigger, and lift itself into an S-shaped attacking posture, ready to strike.

Distribution: South-east Asia.
Habitat: Tropical forest.
Food: Birds and rodents.
Size: 1.6–2.4m (5.3–8ft).
Maturity: Not known.
Breeding: Clutch of 40 eggs.
Life span: Not known.
Status: Common.

Despite its name, none of the red-tailed racer's long, thin body is actually red.

CATS

Different species of cat can be found in many different habitats across the globe. The big cats are distinguished from the small cats by several visible characteristics, although size is not necessarily one of them. Also, the small cats cannot roar. However, cats around the world share the same lithe, sinuous grace and agility, combined with killer instincts and razor-sharp claws.

Tiger

Panthera tigris

The largest and most powerful of the cat family, the tiger's orange and black striped coat helps it to blend into dense undergrowth.

Tigers lack the speed of other cats, but their large and strong hind legs allow them to make leaps of up to 10m (33ft) in length. They have retractable claws, which means that they stay sharp, not getting worn down in day-to-day life. Their foreshortened jaws give them immense crushing power in their bites.

Predominantly solitary, tiger groups usually consist of mothers and their young, which stay with them for two years. Clearly marked home ranges keep accidental meetings and disputes to a minimum, but high levels of social tolerance have been observed, often at kill sites.

Tigers are now critically endangered due to poaching, logging and deforestation.

Distribution: India, Myanmar, Thailand, China and Indonesia.
Habitat: Tropical rainforest, coniferous forest and mangrove swamps.
Food: Hoofed mammals, including pigs, deer and cattle.
Size: 1.9–3.3m (6.25–11ft); 65–300kg (143–660lb).
Maturity: 3–4 years.
Breeding: 3–4 cubs.
Life span: 15 years.
Status: Critically endangered.

Leopard

Panthera pardus

Distribution: Sub-Saharan Africa and southern Asia.
Habitat: Forest, mountains and grassland.
Food: Antelope, deer and rabbits.
Size: 1–1.9m (3.25–6.25ft); 20–90kg (44–200lb).
Maturity: 33 months.
Breeding: 2–4 cubs.
Life span: 10–15 years.
Status: Common.

The leopard is widespread across most of Africa and southern Asia, ranging from open grassland to tropical rainforest and mountain highlands. It is an opportunistic feeder, choosing mainly large hoofed mammals, such as deer and antelope, but will take birds, rabbits and even dung beetles if prey is scarce.

Leopards are well adapted for climbing trees and have been seen hiding the corpses of prey high in the branches to eat later. They hunt mostly during the day to avoid competition with nocturnal lions and hyenas.

Leopards vary greatly in colour depending on their habitat. On the savannah they are usually a sandy ochre, while the high mountain leopards are very dark gold. They tend to have short legs on long bodies, and their fur is covered in black spots or rosettes. Completely black leopards – black panthers – are usually found only in forests, but they do not represent a different subspecies, merely an infrequent mutation.

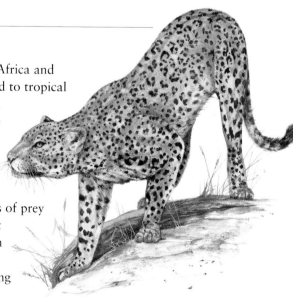

Solitary and nocturnal, the beautiful form of the leopard is rarely seen, even though it often lives in close proximity to humans.

Clouded leopard

Neofelis nebulosa

The clouded leopard is not a close relation of the true leopard, and is believed to be an evolutionary link between the small cats and big cats. This shy and retiring creature has short legs, but a long tail and teeth. It has distinctive cloud-shaped markings on its pelt, which tends to be pale yellow to grey. It also has two large black bars on the back of its neck.

Not much study has been done on clouded leopards due to their shyness of humans. They have been observed running head-first down tree trunks and hanging underneath branches, and will often retire to the trees to digest their meals. They are carnivorous, eating mainly deer, cattle and monkeys. It is thought that most of their hunting is done on the ground.

Like most of the big cats, clouded leopards have no real predators, and their biggest threat comes from human activities. Loss of habitat and pressure of hunting for their beautiful pelts is driving them deeper into the forest. It takes around 25 animals to make a single fur coat. Their bones are also used in traditional medicine.

Distribution: South-east Asia.
Habitat: Deep tropical forest and jungle.
Food: Mainly deer, cattle and monkeys.
Size: 75–90cm (30–36in); 20–30kg (44–66lb).
Maturity: 9 months.
Breeding: 1–5 kittens born in summer.
Life span: 17 years.
Status: Vulnerable.

Elusive and agile, these handsome cats are little known and highly endangered.

Fishing cat (*Felis viverrinus*): 75–85cm (30–34in); 8–14kg (17–31lb)
The fishing cat lives mostly near the mangrove swamps, lakes and marshes of South-east Asia. Unlike a lot of cats, it actually likes water and is able to swim well, in addition to being able to scoop fish from the water with its paws. It is even capable of diving underwater and surfacing below unsuspecting waterfowl.

Flat-headed cat (*Felis planiceps*): 50–65cm (20–26in); 1.5–2kg (3.3–4.4lb)
The flat-headed cat of Indonesia has unusually small and flattened ears and partially webbed toes. It is a naturally good swimmer and hunts mainly for fish and other aquatic animals in mangrove swamps and lakes. In the wild, these cats are predominantly solitary, coming together only to breed. They have short, stumpy legs and, unlike the majority of cat species, cannot fully retract their claws.

Snow leopard (*Panthera uncia*): 1–1.3m (3.25–4.25ft); 25–75kg (55–165lb)
The snow leopard is found on upland steppe and forest in central Asia. It can take prey as large as a yak, but tends to feed on smaller creatures, such as sheep, goats, hares and birds. It is very secretive, hiding away in caves and crevices for much of the day. The snow leopard's luxuriantly thick coat keeps it warm and enables it to live at very high altitudes without suffering from the cold. However, this coat now carries a price: snow leopards are often hunted for their skins, which can be sold for very high prices in Asian countries.

Jungle cat

Felis chaus

Often found living around farms and other human settlements, the jungle cat is a fierce and robust little predator. It has a sandy grey to reddish coat with tabby stripes along its legs, a dark tail tip and black tufts on its large ears. It has good hearing and is very agile, often leaping almost 2m (6.5ft) into the air to catch birds as they fly past.

There is no distinct breeding season, so litters may be born throughout the year. Kittens are generally darker than the adults, with more pronounced stripes. Male jungle cats are often more protective of their young than females, and family groups are common.

This species, along with the African wildcat, was sacred to the ancient Egyptians, who trained them to catch birds – possibly the start of the domestication of cats.

While the jungle cat is common, it shows greater density in natural wetlands than near human habitations, and may be suffering from loss of habitat.

Distribution: Middle East, India and south-east China.
Habitat: Jungles and swamps.
Food: Small mammals, lizards and frogs.
Size: 50–75cm (20–30in); 4–16kg (9–35lb).
Maturity: 18 months.
Breeding: 3–5 young.
Life span: 15 years.
Status: Common.

Jungle cats were worshipped as guardians of the underworld by the ancient Egyptians. Their mummified remains have been found in tombs throughout ancient Egypt.

DOGS

Tough and adaptable, larger species of dog tend to be highly gregarious creatures, living in large and complex social groups called packs. Bonding behaviour in packs is common and important for the stability of relationships between animals of the group. These behaviours include licking, whining and tail-wagging – all traits still readily seen in domesticated dogs.

Asian red dog

Cuon alpinus

The Asian red dog's coat is cinnamon with white patches on the throat and face. The ears are lined with white fur.

The Asian red dog, or dhole, ranges from the alpine forests of Russia to the rainforests as far south as Java, but never lives in open habitats. A highly social animal, it is often seen in packs of around ten animals, sometimes as many as 25. In these packs it hunts large deer and sheep up to ten times as big as itself, and has been seen killing tigers and bears. Larger victims are often partially devoured while still alive. An adult Asian red dog can eat 4kg (8.8lb) of meat in one hour. The animal has a powerful square jaw, enabling it to disembowel its prey easily.

Whether hunting or resting, the Asian red dog leads a well-organized life. The existence of strict hierarchies in packs means that fighting is rare. Females are very sociable and will share their dens with other mothers while giving birth. The males of a pack will help out by hunting and regurgitating food for hungry pups and mothers to eat.

Distribution: Throughout Asia, from India to China.
Habitat: Tropical rainforest and forest steppes.
Food: Small to sizeable mammals, including deer and sheep, occasionally berries and reptiles.
Size: 90cm (36in); 17–21kg (37–46lb).
Maturity: Not known.
Breeding: 8 pups.
Life span: 12 years.
Status: Threatened.

Dingo

Canis dingo

Distribution: Mainland Australia, Burma, Philippines, Indonesia and New Guinea.
Habitat: Forests, woodlands and open arid grasslands.
Food: Rabbits, rodents and marsupials.
Size: 0.7–1.1m (2.25–3.5ft); 8.6–21.5kg (18–47lb).
Maturity: 1 year.
Breeding: 5–6 pups born from June–July.
Life span: 12 years.
Status: Common.

Thought to have arrived in Australia around 4,000 years ago, dingoes are believed to be the ancestors of various true dog breeds. They can be distinguished from dogs that have been descended from wolves by the shape of the skull and through genetic analysis. They usually have ginger coats with white markings.

In Australia, dingoes are widespread and classed as vermin in many states, due to their taste for livestock. When sheep are not available, they will hunt large marsupials, rabbits and lizards. In Asia they tend to exist on rice and other scavenged food, but will also catch small rodents. The Australian dingo is slightly larger than the Asian variety, and their main enemy in both cases are humans. In Australia, bounties are paid for skins and scalps, and in the Asian islands dingoes often make substantial meals for hungry people.

Dingoes were introduced into Australia so long ago that they have evolved into an entirely separate species from domesticated dogs.

Raccoon dog

Nyctereutes procyonoides

The raccoon dog originated in eastern Asia, across northern China and Japan, as well as Siberia and Manchuria. In 1927 it was introduced to eastern Europe for fur-farming and is now seen as far west as the French–German border and northern Finland.

Unusually for a member of the dog family, the raccoon dog is an agile climber, and even the cubs can regularly be seen playing amongst the branches of trees.

Adults form pair bonds and have distinct home ranges. However, these are relatively flexible and they will often roam into other raccoon dog territories. Males and females both help to care for offspring, taking it in turns to guard young while the others hunt for food. They eat whatever they can find, varying with the seasons. Their diet includes fish, amphibians, small mammals, birds, fruit and carrion. They are the only species of dog to "winter sleep". This behaviour resembles hibernation, but doesn't involve a lowering of body temperature.

The raccoon dog looks like a grey and black raccoon, with its characteristic black face mask and brindled greyish body fur.

Distribution: Siberia and north China. Introduced into eastern and central Europe.
Habitat: Damp lowland forest.
Food: Carrion, fruit, fish, frogs and birds. Scavenges food scraps from near human settlements.
Size: 50–60cm (20–24in); 4–10kg (8.8–22lb).
Maturity: 9–11 months.
Breeding: 4–9 young born from April–June.
Life span: 11 years.
Status: Widespread.

Golden jackal

Canis aureus

Distribution: Southern Europe, Middle East, northern Africa and southern Asia.
Habitat: Open savannah and grassland.
Food: Opportunistic feeders.
Size: 0.6–1.1m (2–3.5ft); 7–15kg (15–33lb).
Maturity: Females 11 months; males 2 years.
Breeding: 6–9 pups born after 63 days of gestation.
Life span: 6–8 years.
Status: Common.

Golden jackals can often be seen rummaging around landfill sites near human settlements, looking for tasty refuse.

The golden jackal is widespread, living across southern Europe, the Middle East and south Asia. It is the only jackal that ranges into North Africa, where it was held sacred to the Egyptian god Anubis in ancient times. It usually sports a golden-brown or yellow coat of short, coarse fur and a black-tipped tail.

Golden jackals mate for life and typically raise pups together for about eight years. They live in clearly defined scent-marked territories, often in small family groups. Some offspring remain as helpers, taking care of new-born pups and leaving their mothers free to gather food for their families.

These jackals are found mainly in open grassland terrain. Dominance fights are common, and golden jackals spend more time away from their groups than other social canids. The dominant cubs tend to be the ones that leave their groups, but in unfavourable conditions they will drive the weaker ones from their packs.

Jackals are opportunistic feeders, eating whatever carrion and small mammals they can find, as well as a lot of plant matter. However, golden jackals hunt more than other jackal species, and often come into competition with hyenas and lions, which will try to steal their prey. The jackals eat very quickly, without chewing their food, and will often bury their kills to hide them from other scavengers.

BEARS

Bears are the largest carnivores on land. However, many bears consume high proportions of vegetation in their diets – as much as 95 per cent in some species. Bears have small eyes and ears and large snouts. Their sense of smell is extremely well developed, and this serves them well while foraging and hunting. Bears tend to be exceedingly strong, and a single blow can break a human skull.

Sun bear

Helarctos malayanus

The world's smallest bear rarely stands taller than 1.5m (5ft) and has a short, glossy, black coat with an orange U-shaped marking on its chest. Very little is known about its behaviour and habitat in the wild or, indeed, about just how endangered it might be.

The sun bear is a stocky creature, with large, curved claws for climbing trees and an elongated tongue for eating insects which it finds in the canopy. It will also eat rodents, lizards, honey and the soft insides of palm trees. In more urban areas, it has been seen eating banana crops and refuse. In common with most bears, it is hunted for its medicinal value according to local beliefs, and a combination of hunting and habitat destruction has led to a serious decline in populations of sun bears, with possibly only a few thousand left in the wild.

Barely reaching 1.5m (5ft) in height, the sun bear is the smallest of the bear family.

Distribution: South-east Asia.
Habitat: Lowland tropical rainforest.
Food: Honey, insects and rainforest vegetation.
Size: 1.2–1.5m (4–5ft); 30–60kg (66–132lb).
Maturity: Not known.
Breeding: 2 cubs born after 3.5 months of gestation.
Life span: 25 years.
Status: Endangered.

Sloth bear

Melursus ursinus

The sloth bear is something of an anomaly. In terms of diet, this medium-size bear has more in common with anteaters than other bears. It was named for its ungainly appearance and long, curved claws, but in fact can run faster than a human and is very active and noisy at night. Its preferred food is termites, and to this end it has developed a gap in the front teeth through which it sucks up the insects. It can open and close its nostrils at will, and a hairless patch on the front of the snout protects it against the termites' defensive secretions. A sloth bear at work on a termite nest can be heard a long distance away through the forest. The bear usually lies alone at night. If surprised by humans in the undergrowth, it rears up on its hind legs and brandishes its heavy claws. Sloth bears have been known to injure humans, but they usually run away when disturbed.

The sloth bear is distinguished by its long, black, shaggy hair and its highly dextrous nose.

Distribution: India, Sri Lanka and Bangladesh.
Habitat: Low-altitude dry forest and grassland.
Food: Termites.
Size: 1.5–1.9m (5–6.25ft); 80–140kg (176–308lb).
Maturity: Not known.
Breeding: 2 cubs born each year.
Life span: 40 years.
Status: Vulnerable.

Giant panda

Ailuropoda melanoleuca

One of the most famous and easily recognized animals in the world, the giant panda is also one of the most endangered. Scarcely 1,000 individuals are believed to survive in the wild – in central and south-western China – with another 140 animals in zoos across the world. Habitat loss and poaching are the major dangers, and because the panda has a very slow reproductive cycle, it takes a long time for populations to recover. Females usually give birth to two cubs, one of which survives and stays with the mother for up to three years. In a lifetime, a female may raise only 5–8 cubs.

A panda's diet is almost exclusively bamboo, occasionally supplemented by other grasses and small rodents. Its digestive system is ill-equipped to digest the fibrous bamboo efficiently, so it has to spend most of its days foraging and eating. An elongated wrist bone with a fleshy pad of skin forms a functional but awkward thumb which is used to grasp the stems, and strong teeth then crush them into a more digestible pulp.

Pandas were once thought to be solitary creatures. However, new evidence suggests that small social groups may form outside the breeding season. These well-loved animals are the focus of much detailed research, always aimed at preserving the species. Recent research into in-vitro fertilization of pandas may help in the battle to prevent extinction.

Distribution: Mountain ranges in central China.
Habitat: High-elevation broadleaf forests with bamboo understorey.
Food: Bamboo.
Size: 1.2–1.8m (4–6ft); 110kg (242lb).
Maturity: 4–8 years.
Breeding: 1 cub born every 3–4 years.
Life span: 35 years.
Status: Critically endangered.

Black with white spectacle-like markings around its eyes, the giant panda is instantly recognizable to animal-lovers across the globe.

Asiatic black bear

Ursus thibetanus

The Asiatic black bear has the dubious accolade of being the bear most prized by poachers. Its organs are believed locally to be of particular medicinal potency. Formerly it roamed across most of Asia, from Afghanistan to Japan, but thanks to hunting pressure it is now restricted to small, isolated pockets of high-altitude woodland. It looks similar to the American bears, with black-brown hair and a cream-coloured V-shape across its chest. However, it has larger ears than most other bears.

In the wild, Asiatic black bears are omnivorous, eating small mammals and birds as well as invading bee nests and termite mounds. They are good tree-climbers, despite their short claws, and will spend most of their days aloft. In cooler regions they hibernate, sleeping from November until early April. Females lose nearly half of their body weight during hibernation, so they have to eat well beforehand. In warmer climes the bears often migrate to the lowlands to avoid a winter sleep.

The Asiatic black bear's whitish chest patch gives it the nickname "moon bear".

Distribution: Southern Asia, Russia, Korea and Japan.
Habitat: Deciduous tropical forest and brushland.
Food: Small mammals, insects, honey and fruit.
Size: 1.2–1.9m (4–6.25ft); 90–115kg (198–253lb).
Maturity: 4 years.
Breeding: 1–4 cubs born 6–8 months after mating.
Life span: 25 years.
Status: Endangered.

SMALL CARNIVORES

Small carnivores tend to have long, lithe, sinuous bodies, and are extremely efficient predators, often crucial in the control of rodent populations. Being very adaptable animals, they have been able to carve niches in a wide variety of habitats – from the otters of lakes and seas to the raccoons and martens, some of which now live alongside humans in cities.

Red panda

Ailurus fulgens

The beautiful and elusive red panda is known locally as the "fire fox" in its native country of Nepal.

The red panda is a smaller relative of the more familiar giant panda, although modern biologists are unsure exactly how closely related they are. The red panda is slightly bigger than a large housecat, with a very cat-like face, rusty coloured fur and a long, striped tail. It has partially retractable claws and very tough jaws for chewing the bamboo shoots that are an essential part of its diet. These shoots offer very little nutrition and, like other carnivores, the red panda's digestive system is very short, so it is unable to get the most out of its food. Because of this, the red panda has a low metabolic rate, and chews every mouthful thoroughly. In the wild, red pandas are usually solitary creatures, with clearly defined scent-marked territories. They roam through the forests on the slopes of the Himalayas, which are deciduous hardwoods and rhododendrons with a bamboo understorey. Unfortunately, this fragile ecology and their particular diet means that these beautiful creatures are no longer as widespread as they once were.

Distribution: Himalayas.
Habitat: Highland bamboo forests.
Food: Bamboo leaves, berries, mushrooms and occasionally eggs or young birds.
Size: 510–635mm (20–25in); 3–6kg (6.5–13lb).
Maturity: 18 months.
Breeding: 1–4 cubs per year.
Life span: 14 years.
Status: Lower risk.

Sable

Martes zibellina

Distribution: Siberia and northern Europe.
Habitat: Mountainous forests.
Food: Rodents, birds, fish, nuts and berries.
Size: 35–56cm (14–22in); 0.7–1.8kg (1.5–4lb).
Maturity: 16 months.
Breeding: Litter of 3–4 young born in summer.
Life span: 15 years.
Status: Lower risk.

This carnivore lives in mountainous wooded areas, usually near streams, and an individual may have several dens beneath rocks or large roots. The sable hunts by day or night, roaming across a territory that may be as large as 3,000ha (7,400 acres). Mostly it hunts rodents, but it will also eat small birds, fish, honey and berries.

Sables form individual territories, which are fiercely defended against intruders, but in the mating season the males are more forgiving to passing females. The young are born small and blind during the spring. They open their eyes after around 30 days and are independent by 16 months.

The sable has a luxurious silky coat, usually dark brown or black, and has been hunted for many years. During the 18th century, thousands of animals were trapped for their pelts, and the sable is now raised on farms for the fur industry.

An elegant relative of the pine marten, the sable was almost hunted to extinction for its sumptuous pelage (fur).

Short-clawed otter

Aonyx cinerea

The short-clawed otter is one of the smallest otters in the world and, as its name suggests, has very small claws. These claws are in fact tiny blunt spikes that barely protrude beyond the tips of its paws. Unusually, the feet are webbed only to the last knuckles – not to the ends of the toes, as in other otters. These adaptations mean that it has considerable manual dexterity and sensitivity of touch compared to other otters.

Short-clawed otters tend to catch food using their nimble paws rather than their teeth. They favour mostly crabs and molluscs, and have developed large, broad teeth for cracking the shells. Other food includes frogs and small aquatic mammals. They rarely eat fish, and so there is little competition for food when they live alongside other species of otter.

This Asian otter is highly social, and breeding pairs often stay together for life. Small social groups of up to 12 animals are common, and 12 distinct calls have been recorded. These otters also make rewarding pets, and some have even been trained to catch fish by Malay fishermen.

The short-clawed otter's finger-like front toes make this the most dextrous of the otters. Its front paws are more hand-like than any other otter, allowing it to feel for food in shallow water. It also has a strong tail, making it an excellent swimmer. The short-clawed otter has short dark brown fur with pale markings on its face and chest.

Distribution: South-east Asia, Indonesia and the Philippines.
Habitat: Freshwater wetlands and mangrove swamps.
Food: Crabs, molluscs and frogs.
Size: 450–610mm (18–24in); 1–5kg (2.2–11lb).
Maturity: 2 years.
Breeding: 1–6 young born yearly.
Life span: 20 years.
  Status: Threatened.

Burmese ferret badger (*Melogale personata*): 33–43cm (13–17in); 1–3kg (2.2–6.6lb)
This small, flexible badger has a long bushy tail and white or yellow markings on its cheeks and between its eyes. The ferret badger spends most of its days asleep in burrows, only stirring as night sets in, when it then forages. It eats insects, birds and small mammals and will often climb trees looking for insects, snails and fruit. The Lepcha and Bhotia peoples of northeast India keep Burmese ferret badgers in their homes to control cockroaches and other insect and rodent pests.

Palawan stink badger (*Mydaus marchei*): 32–46cm (12–18in); 3kg (6.6lb)
Found only on the islands of Palawan and Busuanga in the Philippines, the Palawan stink badger lives up to its name, squirting a noxious fluid from its anal glands when it feels threatened. Active by both day and night, it is a slow-moving creature with a stocky body and a long, flexible snout for sniffing out small grubs and worms. Due to its restricted range and secretive habits, little is known about the ecology of this species.

Hog-badger

Artonyx collaris

The hog-badger, or "bear-pig", roams across much of South-east Asia, living mainly in forested areas. It gets its name from its pink, hairless snout and its pig-like feeding behaviour. Rooting in the ground, using its canines and incisor teeth as pick and shovel, it finds small invertebrates and roots, as well as taking fruits and any small mammals that might wander past. Some local peoples have reported it to be a keen fisherman, taking crabs from rivers and streams.

The cubs of this nocturnal animal are playful creatures, but in maturity there is little social interaction. The hog-badger sleeps in deep burrows or caverns under large rocks, and can dig fast enough to escape from some predators – usually leopards or tigers. With strong jaws and sharp claws, it is not an easy meal to catch. In common with other badgers, it also employs a pungent defence mechanism, secreting noxious fumes from its anal glands when in danger.

Distribution: South-east Asia.
Habitat: Jungle and wooded highlands.
Food: Omnivorous: worms, fruit, roots and tubers.
Size: 55–70cm (22–28in); up to 14kg (30lb).
Maturity: 8 months.
Breeding: 2–4 young born in early spring.
Life span: 13 years in captivity.
  Status: Lower risk.

The hog-badger is much like the Eurasian badger in colour and appearance, but with much longer foreclaws and tail. These animals scavenge when food is scarce.

Palm civet (*Paradoxurus hermaphroditus*): 54cm (21in); 3.5kg (7.7lb)
Also known as the toddy cat, this carnivore is found throughout most of South-east Asia, from Timor to India. The palm civet has distinctive markings, with black stripes down its back and small spots on its face. Civets divide their time between the ground and branches, where they feed on fruits, nuts and bulbs. They pick their fruit carefully, leaving less ripe fruit for eating at a later date.

Indian grey mongoose
(*Herpestes edwardsi*): 43cm (17in); 1.5kg (3.3lb)
This mongoose ranges from Arabia to Sri Lanka, living in areas of bush and tall grass. It has a grey-brown coat speckled with flecks of black. The Indian grey mongoose is solitary by nature. A diurnal (day-active) species, it forages on grassland and in open woodland. Its diet is varied, including insects, eggs and small snakes, and it also eats some fruit.

Spotted linsang

Prionodon pardicolor

There are two species of linsang living in Asia, and they are the smallest of the viverrids, the family that includes the cat-like civets and genets. Unlike many other viverrids, they do not produce pungent scents used in defense.

Surprisingly little is known about these carnivores because they are very difficult to observe, especially in their natural forest habitats. They are nocturnal and spend a lot of time in the trees, where they move with sinuous agility through the branches, using their retractable claws to grip on to bark and their long tails for balance. They also forage for food on the ground. During the day, they shelter in nests constructed from sticks and leaves – either in tree hollows or in burrows. Nothing is known about their social behaviour; they are probably territorial like most other small carnivores.

Distribution: Eastern Nepal, southern China and northern Indochina.
Habitat: Forests.
Food: Small mammals, reptiles, birds' eggs and insects.
Size: 30–41cm (12–16in); 598–798g (1.3–1.75lb).
Maturity: 2–3 years.
Breeding: Up to 2 litters of 2–3 young per year.
Life span: 10 years.
Status: Endangered.

Spotted linsangs are sleek, with thick, velvety fur. Their black spots are arranged in rows along their flanks over an orange-buff or pale brown background. Their long tails have dark rings.

Binturong

Arctictis binturong

The binturong is the largest member of the civet family in Asia. Binturongs are nocturnal animals, and spend most of their time up in the trees. Although they are good climbers, they move slowly and carefully through the branches, and have never been observed to make leaps. They are the only carnivores, along with the kinkajou of South America, to have prehensile tails, which they use when climbing.

Binturongs are also capable swimmers, and sometimes dive and hunt for fish. They are easy to domesticate and make affectionate pets. However, surprisingly little is known about these animals in the wild. They can be active both night and day, and although they are usually solitary, one or two adults are sometimes seen together with young. Captive animals make a wide variety of calls. Like so many species, binturongs are declining because of habitat destruction.

The binturong has very long, black, coarse fur, often tipped with grey or buff colours. It has conspicuous tufts of long, straight fur on the backs of its ears, which project well beyond the tips.

Distribution: Sikkim, Indochina, Malaysia, Sumatra, Java and Borneo.
Habitat: Thick forests.
Food: Fish, birds, carrion, fruit, leaves and shoots.
Size: 61–96 cm (24–38in); 9–14 kg (19–30lb).
Maturity: 2.5 years.
Breeding: 2 litters of 1–6 young per year.
Life span: 25 years.
Status: Vulnerable.

RODENTS

The rodents are the largest group of mammals in the world, with over 2,000 species living in a variety of habitats. With the unintentional help of humans, many rodents have greatly expanded their ranges. All rodents have very distinctive incisor teeth, which are used for gnawing. These teeth keep growing throughout the animals' lifetimes to make up for wear and tear from cutting hard materials.

Australian water rat

Hydromys chrysogaster

The Australian water rat, also known as the beaver rat, is one of the few Australian mammals adapted to living in water. It can be found in freshwater rivers, brackish estuaries and saltwater coastal habitats. These rodents are excellent swimmers and hunt for their prey underwater, usually moving along the bottom in search of fish and large aquatic insect larvae. Australian water rats will tackle relatively large prey, including fish up to 30cm (12in) long. They usually carry their prey to feeding sites on logs or rocks.

Unlike many other Australian rodents, the water rat is not strictly nocturnal, and often hunts during the day, though it is most active in the evenings at sunset. This species is very common, and populations may have increased as a result of irrigation projects that have created more suitable watery habitats. However, four closely related species of water rat, which live only in New Guinea or on surrounding islands, are classified as vulnerable.

Distribution: Australia, New Guinea and many surrounding islands.
Habitat: Rivers, streams, swamps and estuaries.
Food: Fish, amphibians and aquatic invertebrates.
Size: 12–35cm (5–14in); 0.3–1.3kg (0.6–2.8lb).
Maturity: 4 months.
Breeding: 1–7 young per litter; 2–3 litters per year.
Life span: 6 years.
Status: Common.

The Australian water rat has partially webbed hind feet used for paddling, and thick, sleek, seal-like fur.

Golden hamster

Mesocricetus auratus

A number of species of small rodent are commonly referred to as hamsters, all of which belong to the same subfamily, the *Cricetinae*. Although wild populations of golden hamsters live in only a small area of Syria, people have introduced them into many other parts of the world. Golden hamsters are easy to keep, and have become very popular pets.

In the wild, this species lives in burrows in dry, rocky habitats. Burrows can be quite large and often comprise a number of chambers and entrances. Hamsters live alone in their burrows and are usually very aggressive towards one another. Indeed, captive hamsters must be kept apart in order to prevent them from killing one another.

Golden hamsters have huge cheek pouches for carrying food back to their burrows. If their young are threatened, female hamsters have been known to place as many as 12 baby hamsters in their cheek pouches. Although golden hamsters are common household pets and the third most common laboratory animal after rats and mice, the wild population inhabits such a small area that the species is listed as endangered.

Distribution: Wild population restricted to Syria in the region of Aleppo.
Habitat: Semi-desert, rocky hills and scrubland.
Food: Wide variety of plant material, including leaves, seeds and fruit. Also known to eat meat.
Size: 15–20cm (6–8in); 97–200g (0.2–0.4lb).
Maturity: 2 months.
Breeding: 2–5 litters of 2–16 young produced annually.
Life span: 3 years.
Status: Endangered.

Wild specimens have rich golden-brown fur on the upper side, but selective breeding in captive populations has produced a wide range of colours and fur types.

Flying squirrel

Petaurista elegans

Distribution: Nepal, Burma, South-east Asia, southern China, Sumatra, Java, Borneo and some surrounding islands.
Habitat: Mountain and hill forests.
Food: Fruit, nuts, leaves, shoots and maybe also insects and their larvae.
Size: 30.5–58.5cm (12–23in); 1–2.5kg (2.2–11lb).
Maturity: 1 year.
Breeding: 1 young.
Life span: 15 years.
Status: Common.

This animal, also known as the spotted giant flying squirrel, is one of the largest of over 40 species of flying squirrel found in the world. There are ten species of giant flying squirrel, all of which live in high-altitude forests in eastern Asia, at elevations usually exceeding 900m (3,000ft). In the Himalayas, the spotted giant flying squirrel lives as high up as 4,000m (13,000ft).

During the day, these animals hide in hollow tree trunks or in the thick foliage of high branches. At dusk, they start moving about in search of food. Usually, they move through the treetops in the same way as common tree squirrels, jumping from branch to branch. However, when they come to gaps in the trees too wide to cross by jumping, they leap from high branches and glide across the gaps on the broad webs of skin stretched between their limbs and tail. They have been known to make leaps as long as 450m (1,500ft). Although these animals don't actually fly, they are quite skilful in the air, making several banking turns in one leap and even riding rising air currents.

Giant flying squirrels have very long bushy "bottlebrush" tails, which are not flattened like those of some other species of flying squirrel. The tail acts as a rudder, both when the squirrel is leaping and when gliding through the air.

Giant squirrel

Ratufa indica

Distribution: Peninsular India.
Habitat: Forests.
Food: Fruit, nuts, bark, insects and birds' eggs.
Size: 25–46cm (10–18in); 1.5–3kg (3.3–6.6lb).
Maturity: 2 years.
Breeding: 1–5 young.
Life span: 20 years.
Status: Vulnerable.

The giant squirrel has dark brown fur on its upper side and buff-coloured underparts. It has particularly broad hands and long claws to help it grasp branches and climb.

This is the world's largest species of tree squirrel, with a head to tail length of up to 90cm (36in) and a weight of nearly 3kg (6.6lb). Giant squirrels spend most of their time in the treetops and are very agile, moving rapidly through the forest canopy. These squirrels are active during the day, and shelter in holes in trees at night.

During the breeding season, they construct large spherical nests from sticks and leaves, where the females give birth and care for their young. Giant squirrels live alone or in pairs, and are thought to have relatively small home ranges, centred on their nest sites. The squirrels make low "churring" calls when content, and loud chattering calls when alarmed. They are very wary and will dash into thick forest at the least indication of danger. As a result, they are very difficult to study, and relatively little is known about the species. They are now endangered due to the destruction of their habitats.

Beautiful squirrel
(*Callosciurius erythraeus*): 13–28cm (5–11in); 150–500g (0.3–1.1lb)
This is one of 15 species of so-called beautiful squirrels, found in the forests of India and eastern Asia. The squirrels are among the most brightly coloured of all mammal species, and often have a distinctive three-colour pattern. This species has an olive-coloured upper side and a rich red-brown underside.

Long-eared jerboa
(*Euchoreutes naso*): 7–9cm (2.8–3.6in); 30–50g (0.06–0.1lb)
This curious rodent from Mongolia and northern China has exceptionally large ears – longer than its head. Like other jerboas, it hops around on long back legs, like a miniature kangaroo, keeping its balance with the aid of a long, tufted tail. This little-known jerboa is now thought to have become endangered due to the loss of its habitat.

BATS

Bats are the second largest group of mammals – with 977 species – and the only mammals capable of true flight. Bats live in all temperate and tropical regions around the world and, because of their ability to fly, they have even colonized remote oceanic islands. Bat species are not evenly distributed around the world. For example, there are over 100 species of fruit bat living around Indonesia and New Guinea.

Flying fox

Pteropus edulis

Flying foxes are fruit bats with fox-like muzzles and ears, and large eyes.

This is one of 60 species of flying fox distributed over South-east Asia and Australia, as well as many islands of the western Pacific and Indian Ocean.

These bats spend their days roosting high up in tall trees, only emerging from the forest canopy at dusk, flying off in search of food. Fruit bats take off by dropping from their branches, so taller trees are preferred because they provide good launch sites. Sometimes fruit bats roost together in huge numbers. In the Philippines this species used to gather in groups of 100,000 before hunting and habitat loss drastically reduced its numbers.

Fruit bats are hunted for their meat, which is considered a delicacy in many countries. In some countries, fruit growers consider them pests. Isolated populations are vulnerable to local catastrophes such as tropical storms. In the last 50 years, two species have become extinct, and many more are seriously endangered.

Distribution: South-east Asia, Indonesia, the Philippines and the Nicobar and Andaman Islands.
Habitat: Lowland tropical forest and surrounding agricultural land with trees.
Food: Fruit, flowers, nectar and pollen.
Size: 1–1.8m (3.25–6ft); 0.6–1.6kg (1.3–3.5lb).
Maturity: 2 years.
Breeding: Single pup born annually.
Life span: 30 years in captivity.
Status: Common.

Schneider's leaf-nosed bat

Hipposideros speoris

There are 69 species of leaf-nosed bat which live in tropical and subtropical parts of Africa, Asia and Australia. These bats have strangely shaped noses which help them to focus their ultrasonic vocalizations – calls emitted from the nostrils and used in echolocation. These bats can locate flying insect prey and determine how fast they are moving, as well as detecting obstacles, such as branches, from the type of echoes returning from their calls.

Different echolocating bats make different types of call, depending on the kind of habitat in which they live. Schneider's leaf-nosed bats hunt insects in a particularly wide range of habitat types, including inside forests and the edges of more open areas. Although they can hunt and navigate in pitch darkness, these leaf-nosed bats often hunt around dusk and dawn when many species of swarming insects are most active. However, during these times they are more vulnerable to predators such as birds of prey. Although Schneider's leaf-nosed bats are fairly common throughout their range, 13 other species from the same genus are either declining in number or in danger of extinction.

Distribution: Southern India and Sri Lanka.
Habitat: Roost in hollow trees, caves and buildings.
Food: Insects taken in flight.
Size: 4–9cm (1.6–3.6in); 20–50g (0.04–0.11lb).
Maturity: 18–19 months.
Breeding: Single young born annually.
Life span: 12 years.
Status: Not known.

The muzzles of these bats have strange leaf-like growths of skin, consisting of frontal horseshoe-shaped parts and leaf-like structures running across their centres. These structures focus the ultrasonic calls that help them catch insect prey.

Short-tailed bat (*Mystacina tuberculata*): 60–68mm (2.4–2.7in); 12–15g (0.02–0.03lb)
This is one of two species of bat found in New Zealand – the only native mammals to live there. Like many forest bats, this species roosts in tree hollows but, unusually among bats, it is a ground forager and can move with great agility along the forest floor, where it preys on large invertebrates.

Bulmer's fruit bat (*Aproteles bulmerae*): 24cm (9.5in); 600g (1.3lb)
This rare fruit bat was first known from fossil remains in central Papua New Guinea, and living specimens were later discovered in the 1970s. Unlike most species of fruit bat, it roosts in caves rather than in trees. This species was thought to have gone extinct until some individuals were rediscovered in the early 1990s.

Pygmy fruit bat (*Aethalops alecto*): 65–73mm (2.5–2.9in); 19g (0.04lb)
This is the smallest of the fruit bats of Asia, and lives in high-altitude forests on the mountains of Malaysia and Indonesia. This nocturnal bat feeds on fruit and pollen, and during the day it roosts alone or in twos or threes. The species is threatened by the deforestation of its habitat.

Common blossom bat

Syconycteris australis

These small fruit bats are members of a group of similar bats, all of which have long, slender tongues with brush-like projections for picking up pollen and nectar from flowers. Although blossom bats usually roost alone, they may form "camps" like those of flying foxes, roosting together in groups at certain times of the year.

Some blossom bats stay in the same areas, apparently trying to defend their flowers from other bats, while others move freely between different foraging areas. Like many fruit bats, this species has large eyes, well adapted for seeing in low light.

At night, blossom bats congregate in areas with many flowers, and large numbers can be seen flying around some species of tree when they are in bloom. Perhaps surprisingly, relatively few blossom bats attempt to defend rich nectar sources because of the increased disturbance from the large number of bats attracted to such sites.

Distribution: North-eastern Australia, New Guinea and several island groups of Indonesia and the western Pacific.
Habitat: Wide range of forest types.
Food: Nectar, pollen and occasionally fruit.
Size: 5–7.5cm (2–3in); 11.5–25g (0.02–0.05lb).
Maturity: Not known; probably around 1 year.
Breeding: 1 young or very occasionally twins born.
Life span: Not known.
Status: Common.

Common blossom bats are one of the smallest fruit bats in the world. They have red-brown or grey-brown fur on top, and pale undersides. Like some species of fruit bat, they are important pollinators. There is concern that their numbers are declining in Australia.

Colugo

Cynocephalus variegatus

The colugo is also known as the flying lemur, but this name is somewhat misleading because these animals are not lemurs, and although they can glide almost horizontally for 100m (330ft) or more, they are not capable of true flight either.

Colugos have very well developed gliding membranes which, unlike those of other gliding mammals, stretch all the way to the tips of their fingers, toes and tails. As their wide eyes suggest, these animals are largely nocturnal, sleeping in holes or hollows high up in trees during the daytime. At dusk they leave their shelters, climbing up to find good take-off positions, and then glide away in search of food.

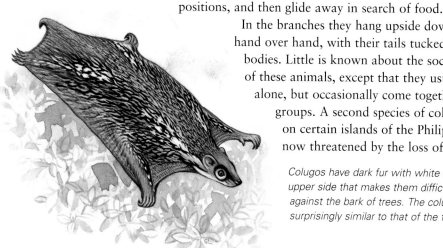

In the branches they hang upside down, moving hand over hand, with their tails tucked under their bodies. Little is known about the social behaviour of these animals, except that they usually live alone, but occasionally come together in small groups. A second species of colugo only lives on certain islands of the Philippines, and is now threatened by the loss of its habitat.

Colugos have dark fur with white spots on their upper side that makes them difficult to see against the bark of trees. The colugo's head is surprisingly similar to that of the flying fox.

Distribution: South-east Asia and the East Indies.
Habitat: Primary and secondary forest, coconut and rubber plantations.
Food: Young leaves and perhaps also fruit, flowers and buds.
Size: 32–42cm (12–17in); 1–1.75kg (2.2–3.8lb).
Maturity: Not certain, but probably around 3 years.
Breeding: 1 or 2 young are reared annually.
Life span: 15 years.
Status: Not known.

ELEPHANTS AND RELATIVES

The first ancestors of the elephants originated in Africa around 40 million years ago. From this prehistoric family, two different groups evolved: the mastodons and the modern elephant family. The relatives of the Asian elephant appeared in Africa, and then spread into Asia and Europe. Elephants have few relatives. They are thought to share ancestry with hyraxes and sea cows – dugongs and manatees.

Asian elephant

Elephas maximas

Asian elephants can be distinguished from their African cousins by their smaller ears, sloping backs and by having only one rather than two finger-like projections on the ends of their trunks.

The Asian elephant is one of the world's largest land animals, second only to its close relative, the African elephant. Asian elephants live in small groups of 15–40 individuals, consisting of related females and their young, led by old matriarchs – head females.

Asian elephants are intelligent animals, and they have been reported to be able to use tools. For example, they sometimes use sticks held in their trunk to scratch themselves or to swat insects. They have been domesticated for many thousands of years and have been used as draught and war animals up to modern times. However, the Asian elephant can be an agricultural pest, eating up to 150kg (330lb) of crops per day.

Distribution: Originally found in the Middle East, India, southern China, South-east Asia, Java, Sumatra, Borneo and Sri Lanka.
Habitat: Tropical forest, open woodland and grassland.
Food: Grass, leaves, roots, stems, fruit and other crops.
Size: 5.5–6.4m (18–21ft); 2,720–6,700kg (6,000–14,750lb).
Maturity: 9 years.
Breeding: 1 calf born every 2–8 years.
Life span: 80 years.
Status: Endangered.

Dugong

Dugong dugon

The dugong is a very distant relative of the elephant, and is placed in its own order – the *Sirenia* – along with the manatees. Dugongs live in shallow coastal regions where the sea grass on which they feed is abundant. They rarely make long-distance migrations, though in some places they make daily movements from feeding areas to resting sites in deeper water.

Dugongs have unusually shaped mouths, with overhanging upper lips that are specially designed for cropping sea grasses. They can dive for up to three minutes before coming up to breathe, and will swim at up to 20kph (12mph) if pursued.

The young are born underwater, after which they ride on their mothers' backs, breathing when the females come to the surface.

Sharks attack dugongs, but groups of dugongs will gang up on them and ram them with their heads. Orcas (killer whales) have also been known to attack dugongs, but by far their greatest enemy is human beings, who have hunted them extensively for their meat, hides and ivory.

Dugongs have thick, smooth hides, usually dull grey-brown in colour. Unlike the closely related manatees, which have rounded paddle-shaped tails, dugongs have fluked tails like those of whales and dolphins.

Distribution: Coasts of tropical East Africa, tropical Asia and Oceania.
Habitat: Shallow coastal waters.
Food: Sea grass and occasionally algae and crustaceans.
Size: 2.4–2.7m (8–9ft); 230–908kg (500–2,000lb).
Maturity: 9–15 years.
Breeding: 1 or 2 young born after about 1 year's gestation, every 3–7 years.
Life span: 70 years.
Status: Vulnerable.

HOOFED ANIMALS

Hoofed animals are separated into two main groups by the shape of their hooves. The Perissodactyla, *which includes the horses, zebras, tapirs and rhinos, have hooves with an odd number of toes, with most of the animal's weight supported on the middle toe. The* Artiodactyla, *in contrast, have an even number of toes, and include animals such as deer, cattle, antelope, pigs, camels, giraffes and hippopotamuses.*

Muntjac

Muntiacus reevesi

Only the males of these dainty deer carry antlers, and these rarely exceed 15cm (6in) in length. However, the male muntjac's main weapons are its long upper canine teeth, which curve outwards from its lips like tusks. The teeth are usually used in fights to settle territorial disputes with other males, but can also inflict savage injuries on dogs and other attacking animals.

As well as the protruding canine teeth that are also found in a few other species of deer, the lower portions of a male muntjac's antlers are covered in fur. Instead of antlers, females have small bony knobs with a coating of fur.

Male muntjacs defend territories of around 20ha (50 acres) in size, usually close to water, and try to encompass as many of the smaller female territories as possible. When they are nervous, muntjacs make a call that sounds like a dog barking, giving muntjacs their other common name, "barking deer". These calls are probably aimed at predators that use ambush tactics to catch their prey. By calling, the muntjacs let the predators know that they have been spotted, and therefore are unlikely to succeed in their ambushes. Muntjacs were introduced into Britain in 1901, and feral populations have established themselves well.

Distribution: Southern China and Taiwan. Introduced to Britain.
Habitat: Forest and farmland surrounded by vegetation.
Food: Grass, tender leaves and shoots.
Size: 0.6–1.4m (2–4.75ft); 14–33kg (30–73lb).
Maturity: Females 1 year; probably older for males.
Breeding: Single fawn born each year.
Life span: 15 years.
Status: Common.

Sambar

Cervus unicolor

This large Asian deer has particularly stout six–pointed antlers that can measure as much as 1m (3.25ft) along the outside curve. Sambars are very cunning and can live close to human habitation.

The sambar is one of ten closely related species of deer, all in the genus *Cervus*, which includes the European and American red deer and the oriental sika deer. Like red deer, male sambars fight during the mating season. Victorious males set up small territories marked with scent, and mate with any females that pass through. In some nature reserves in India, sambars are important prey animals for dholes (Asian red dogs), Asiatic lions and tigers. Although they are an important prey for some rare predators, sambars can also be troublesome to nature reserve managers because they sometimes kill large trees by stripping their bark, leading to considerable habitat destruction. In some countries where sambars have been introduced, such as New Zealand, they are hunted for sport.

Distribution: India to south-eastern China, Malay peninsula, Sri Lanka, Taiwan, Hainan, Sumatra, Borneo and surrounding islands. Introduced into Australia, New Zealand, California, Texas and Florida.
Habitat: Forests.
Food: Leaves, buds, berries, twigs, grass and fallen fruit.
Size: 1.6–2.5m (5.25–8.25ft); 109–260kg (240–570lb).
Maturity: Females 2 years; first mating older in males.
Breeding: Single fawn.
Life span: 25 years.
Status: Abundant.

Przewalski's horse

Equus przewalskii

Distribution: Altai Mountains of Mongolia.
Habitat: Grassy plains and deserts.
Food: Grass, leaves, twigs, buds and fruit.
Size: 2.2–2.8m (7.25–9.25ft); 200–300kg (440–660lb).
Maturity: Females 4 years; males 5 years.
Breeding: Single foal born every 3 years.
Life span: 38 years.
Status: Extinct in the wild.

Some scientists consider this horse as simply a subspecies of the domestic horse – *Equus caballus* – rather than a separate species. Most certainly, Przewalski's horse represents a unique animal, either as the last remaining truly wild variety of domestic horse, or as a unique species. In fact, Przewalski's horse, named after the Russian explorer who discovered it in 1879, has very similar habits to feral populations of domestic horses.

It lives in groups of around ten individuals, consisting of females with their young and a single dominant male or stallion. The stallion may lead his harem of females and young for many years, taking great care to protect them from rival males and other dangers such as predators.

These horses spend their days in dry areas, and in the evenings move to wetter areas with better grazing and drinking water. In some parts they make seasonal migrations to track the rainfall and the richest grazing sites. The horses have not been seen in the wild since 1968, and it is feared that they are now extinct in their former range. However, the species survives in zoos.

Przewalski's horses are slightly smaller than domestic horses, with light reddish-brown-coloured upper parts, pale flanks and white undersides. Plans have been made to reintroduce these horses to their former range.

Sika deer (*Cervus nippon*): 0.9–1.4m (3–4.75ft); 80kg (176lb)
This species of deer lives in eastern Asia, including most of China and Japan. Like red deer, male sika deer fight during the rutting season in order to control harems of up to 14 females. Sika deer are very vocal, and can produce at least ten different sounds, ranging from soft whistles to loud screams.

Giant muntjac (*Megamuntiacus vuquangensis*): 96cm (38in); 34–50kg (75–110lb)
This animal was unknown to science until 1994, when descriptions of a few individuals found in Vietnam were first published. Like other species of muntjac deer, the males have sharp elongated canines used in fighting. Giant muntjacs are hunted for their meat, and may also be suffering from habitat loss.

Four-horned antelope or **chousingha** (*Tetracerus quadricornis*): 0.8–1m (3–3.25ft); 17–21kg (15–46lb)
This small antelope lives in Nepal and India in open forests, usually close to water. The males are unique among the *Bovidae* – the family including antelope, cattle and goats – in that they have four horns. These horns are smooth and cone-shaped, and have made this species sought after by trophy hunters.

Babirusa

Babyrousa babyrussa

There are probably little more than 4,000 of these wild pigs roaming the Indonesian forests of Sulawesi and its surrounding islands.

Babirusas are most active during daylight hours. Males usually live alone, while females and young go about in groups of about a dozen. The tusks are not well designed for combat, and although males are aggressive when they meet, they rarely use their tusks in fighting. Females do not usually fight among themselves, but they sometimes attack by trying to bite the front legs of their opponents. Babirusas are fast runners and good swimmers. They sometimes swim out to sea to reach small islands.

Distribution: Sulawesi, Indonesia, and some nearby islands.
Habitat: Damp, tropical forests.
Food: Leaves and fallen fruit.
Size: 0.9–1.1m (3ft); up to 100kg (220lb).
Maturity: 1 year.
Breeding: 2 litters of 1 or 2 young produced annually.
Life span: 24 years.
Status: Vulnerable.

Unlike any other species of wild pig, whose tusks grow from the sides of the jaw, the upper tusks of the babirusa grow upwards through the muzzle. These pigs often visit salt licks, where scientists observe their behaviour.

Bactrian camel

Camelus bactrianus

Archaeological evidence has revealed that bactrian, or two-humped, camels were first domesticated around 2500 BC. It used to be believed that the one-humped, or dromedary, camel evolved from domesticated bactrian camels. However, 3,000-year-old rock drawings in the Arabian peninsula show horsemen hunting dromedaries, and 7,000-year-old dromedary remains suggest that one-humped camels had wild-living ancestors.

Bactrian camels and dromedaries can interbreed, and their young have either a long single hump with a slight indentation, or a large hump and a small hump. Camel humps contain fat deposits that provide the animal with energy when there is no available food.

Over the last hundred years there has been a great reduction in wild bactrian camel numbers. Although there are around a million domesticated animals, only 1,000 or so wild ones remain, and these are affected by hunting and herders who prevent them from reaching water holes. Camels have many adaptations to cold and heat, and can survive for long periods without water. They have very few sweat glands, so lose little liquid through their skins. They are able to drink brackish water, which would make other animals sick.

Wild bactrian camels are quite different in appearance from domestic ones, and have much lighter coats, shorter fur, leaner bodies and smaller humps.

Distribution: Wild populations now only found in Mongolia and north-western China, but domestic animals found in many parts of central Asia and beyond.
Habitat: Dry steppes and semi-deserts.
Food: Unpalatable plants and carrion.
Size: 2.2–3.4m (7.25–11.25ft); 300–690kg (660–1,500lb).
Maturity: Reach maturity after more than 3 years.
Breeding: Single calf produced every 2 years. Occasionally twins are born.
Life span: Probably around 20 years in the wild.
Status: Endangered.

Indian rhinoceros

Rhinoceros unicornis

This is the largest of three different species of rhinoceros living in Asia. Apart from mothers with their calves, these rhinos live alone, though occasionally several come together at muddy wallows or good grazing areas. They live in home ranges of 2–8sq km (0.75–3sq miles), which overlap with those of other rhinos. Meetings between neighbouring rhinos are often aggressive affairs. Although Indian rhinoceroses usually flee when disturbed, they sometimes charge at humans. Females with calves are particularly dangerous, and several fatal attacks are recorded every year in Nepal and India.

This species used to be common in north-western India and Pakistan until around 1600, when large areas of lush lowland grasslands were turned into farmland. As well as losing much of their prime habitat, rhinos came into conflict with farmers and sportsmen. By the early 1900s the species was close to extinction. International law now protects these rhinos, and their numbers have increased. There are now around 2,000 individuals in the wild.

Distribution: Originally found in northern Pakistan, Nepal, northern India, northern Bangladesh and Assam.
Habitat: Grasslands, swamps, forests and farmland.
Food: Grass, leaves, fruit, twigs and crops.
Size: 3.1–3.8m (10–12.5ft); 1,600–2,200kg (3,500–4,800lb).
Maturity: 7–10 years.
Breeding: Single calf born every 3–5 years.
Life span: 40 years.
Status: Endangered.

Both males and females have horns that can grow up to 53cm (21in). The skin is greatly folded, and is covered in rivet-like knobs that make it look like armour.

Yak

Bos grunniens

Yaks are closely related to cattle, and have been domesticated since around 1000 BC. There are nearly 13 million domestic yaks around the high plateaux of Central Asia, where they are used as draught animals and for milk, meat and wool production.

Yaks are well suited to cold, high-altitude conditions, being powerful but docile. There are probably only a few thousand wild yaks, ranging from eastern Kashmir along the Tibetan–Chinese border into the Qinghai province of China.

Yaks spend the short summers feeding in sparse alpine meadows, and in winter they descend into the valleys. Females usually live together in large herds that used to be thousands strong when wild yaks were more abundant. Males live alone or in small groups of less than 12 animals until the breeding season, when they join the herds to fight over females.

Yaks have stocky bodies and very long black-brown woolly fur that almost reaches to the ground, to help keep them warm. Wild yaks are larger and have stronger horns than domestic yaks.

Distribution: Originally found in highlands of Siberia, Nepal, Tibet, western China and adjacent areas around the Himalayas.
Habitat: Highland and mountainous steppes.
Food: Grass, herbs and lichens.
Size: 3.3m (11ft); 400–1,000kg (880–2,200lb).
Maturity: 6–8 years.
Breeding: Single calf every 2 years.
Life span: 25 years.
Status: Endangered.

Gaur (*Bos gaurus*): 2.5–3.3m (8–11ft); 650–1,000kg (1,400–2,200lb)
The gaur is closely related to domestic cattle, and is characterized by its short red-brown or black-brown fur, with white lower legs. This species lives from Nepal and India to South-east Asia, including the Malay peninsula, inhabiting forested hills. In some places the gaur is diurnal, in others it is nocturnal, depending on the level of persecution by humans.

Mountain anoa (*Bubalus quarlesi*): 1.6–1.7m (5–5.5ft); 150–300kg (330–660lb)
There are thought to be two species of anoa, both restricted to the island of Sulawesi. Mountain anoas live in dense forests and are very secretive. Usually anoas will flee when disturbed, but occasionally they have been known to charge. These animals have a very restricted range and have suffered from excessive hunting and habitat loss.

Axis deer or **chital** (*Axis axis*): 1–1.7m (3.25–5.5ft); 27–110kg (60–240lb)
Found in India, Nepal and Sri Lanka, these deer are characterized by their slender body shape. At certain times of the year they have white spots on their fawn-coloured fur. They live mostly in open habitats, presumably so that they can see approaching predators. Indeed, axis deer are the favourite prey of tigers in many parts of India.

Nilgai

Boselaphus tragocamelus

Nilgai used to be widespread across the Middle East but in recent years they have disappeared from Bangladesh, though they continue to survive in India, Pakistan and Nepal, and have been introduced into Texas.

During the mating season, males compete for territories, usually by aggressive displays or ritualized fighting, involving pushing each other with their necks. Occasionally, proper fighting occurs, and they drop to their knees, lunging at each other with their short stabbing horns. The winners have the opportunity to gather harems of up to ten females.

Distribution: Eastern Pakistan, India, and Nepal. Introduced into Texas.
Habitat: Forests, jungles and occasionally open grasslands.
Food: Grass, leaves, twigs, and will also eat fruit and sugar cane.
Size: 1.8–2.1m (6–7ft); up to 300kg (660lb).
Maturity: Females 2 years; males 5 years.
Breeding: Occasionally 3 calves born at a time; twins very frequent.
Life span: 10 years.
Status: Conservation-dependent.

Only the male nilgai have horns, but both sexes possess manes on their necks. The wiry coat is reddish-brown in males and a paler colour in females.

Serow (*Capricornis sumatraensis*): 1.4–1.8m (4.75–6ft); 50–140kg (110–308lb)
This is the most widespread of three species of serow living in large parts of China, the Himalayan region, South-east Asia and parts of Indonesia. The other two species are restricted to the islands of Taiwan and the Japanese archipelago. Serows are related to goats, and similarly live on rocky slopes. Both males and females possess sharp horns rarely more than 25cm (10in) long, which can inflict deadly injuries to hunting dogs.

Sao la (*Pseudoryx nghetinhensis*): 1.5–2m (5–6.5ft); 100kg (220lb)
Despite its large size and striking colours, the sao la, with its long, straight, smooth horns and striped facial pattern, was unknown to science until 1992. This animal was first identified from several sets of horns obtained from hunters in a nature reserve on the Laos-Vietnam border. Unfortunately, shortly after its discovery it was classified as endangered, and is threatened by poaching and habitat loss.

Tibetan antelope or **chiru** (*Pantholops hodgsoni*): 1.3–1.4m (4.25–4.5ft); 25–55kg (55–120lb)
This animal is actually more closely related to goats than to true antelopes, and lives on the high plateau of Tibet, Kashmir and north-central China. The males have slender, ridged horns up to 70cm (28in) long, rising straight upwards. They use these sabre-like horns when fighting for control of females during the rutting season, sometimes inflicting fatal injuries.

Himalayan tahr

Hemitragus jemlahicus

Himalayan tahrs are closely related to goats. There are two other species of tahr: the Arabian tahr, which is only found in Oman, and the Niligri tahr, which lives around the Niligri Hills in southern India.

Both male and female Himalayan tahrs have luxuriant manes over their necks and shoulders, presumably as a defence against the cold mountain air. Tahrs are very wary, and at the least alarm they will scamper off over rocks and through forest, moving easily over the steep terrain.

Tahrs live in herds usually consisting of around ten individuals. During the mating season, males lock horns and wrestle one another in order to win the right to mate with females.

Competition with domestic goats and hunting has begun to reduce all three species of tahr. Arabian and Niligri tahrs are now officially classified as endangered by the IUCN.

Distribution: The Himalayas, from Kashmir to western Bhutan. Introduced into New Zealand.
Habitat: Forested hills and mountains.
Food: Mostly grass.
Size: 0.9–1.4m (3–4.75ft); 50–100kg (110–220lb).
Maturity: Females around 1.5 years; probably older for males.
Breeding: 1 or 2 young at a time.
Life span: 21 years.
Status: Vulnerable.

Himalayan tahrs resemble goats, but unlike goats, males do not sport goatee beards and do not have twisted horns.

Saiga

Saiga tatarica

Saigas are related to goats and sheep, and around two million years ago, in the Pleistocene epoch, these animals were very abundant, ranging right across Europe and Asia from Britain to Alaska. Nowadays, saigas are not so common. They have been exterminated from some parts of their range, such as Crimea, by over-hunting and habitat loss.

Saigas are nomadic, constantly moving about in search of food. In autumn, many saiga populations migrate south to warmer climes and better feeding grounds, returning north in the following spring. Saigas move rapidly, covering 80–120km (50–75 miles) per day during their migrations.

Adult saigas can run at up to 80kph (50mph), and even two-day-old saiga kids can outrun humans. In winter, after the migration is over, males try to herd females into their territories, and successful males gather between 5–15 females. Competition is intense, and fights between saiga males are sometimes fatal.

The saiga has an unusually shaped nose with large, downward-pointing nostrils. The strange internal structure of this nose is only found in one other type of mammal – whales. It is thought to warm inhaled air in cold winters, and to help cool the saiga during hot summers.

Distribution: Western Ukraine to western Mongolia.
Habitat: Steppes and dry grassy plains.
Food: Grass, herbs and shrubs.
Size: 1–1.4m (3.25–4.75ft); 26–69kg (57–150lb).
Maturity: 12–20 months.
Breeding: 1–2 calves.
Life span: 10–12 years in the wild.
Status: Vulnerable.

Water buffalo

Bubalus bubalis

Distribution: Nepal, India, Sri Lanka, South-east Asia, Malay peninsula, Borneo, Java and Sumatra. Introduced into many other areas, including South America, Europe, Australia and Hawaii.
Habitat: Wet grasslands, swamps and lush river valleys.
Food: Green grass and water vegetation.
Size: 2.4–3m (8–10ft); 700–1,200kg (1,540–2,640lb).
Maturity: 18 months.
Breeding: Single calf every 2 years.
Life span: 25 years.
Status: Endangered in the wild; domestic population is common.

These large animals were domesticated as long ago as 5000 BC in southern China, 3000 BC in the Indus valley and 2000 BC in other parts of the Middle East. There are now estimated to be around 150 million domestic water buffalo around the world. So useful are these beasts that they have been introduced into many new areas, including Australia, South America, southern Europe and Hawaii.

Domestic water buffalo are very docile, produce excellent milk and meat, and are strong and easily managed work animals. In some areas, they form an important part of agricultural economies and ecosystems by providing a reliable and easily maintained source of power and by conserving wallowing sites that harbour a wide diversity of animal and plant life.

Wild water buffalo from original undomesticated stock are very rare, and very few populations remain. These wild buffalo live in herds of up to 30 individuals, consisting of females and their young, led by elder females. During the dry season males live in bachelor groups away from the females, and in the wet season the dominant males enter herds to mate. The female herd leaders remain in charge of their groups even when bulls are present, and after mating the males are driven away.

Wild water buffalo have the longest spread of horns of any cattle – up to 2m (6.5ft), measured along the outside edge.

Markhor

Capra falconeri

The markhor is one of seven species of wild goat, all of which live in mountainous regions of Europe and Asia. This species lives in certain mountain ranges of Central Asia. As a protection against the cold continental winters, markhors have thick white and grey coats and long shaggy manes covering their necks and shoulders. In summer, markhors lose their winter coats, which are replaced by shorter red-grey coats, and they move higher up the mountains to feed on the rich spring growth.

Male markhors are characterized by their long corkscrew-shaped horns that can reach up to 1.6m (5.25ft) in length. Females also have horns, but they rarely grow longer than 25cm (10in).

Females live in small herds, usually comprising about nine individuals, but occasionally up to 100. Males are solitary, only joining the female herds in the mating season. During this time, they fight one another for females, lunging at one another and locking horns.

Markhors are sought by trophy hunters and are also killed for their meat and hides. There are three subspecies, the rarest of which numbers only around 700 individuals. The political instability that has plagued some of the countries in the markhor's range has made it difficult to control illegal poaching.

Distribution: Mountainous areas of central Asian countries, including Turkmenistan, Uzbekistan, Tajikistan, Afghanistan, Pakistan and Kashmir.
Habitat: Mountainous areas close to the treeline, rocky areas, dry country and steep meadows.
Food: Grass, leaves and twigs.
Size: 1.4–1.8m (4.75–6ft); 32–110kg (70–240lb).
Maturity: 2.5 years.
Breeding: 1 or 2 young born at a time.
Life span: 12 years.
Status: Endangered.

TREE SHREWS

Tree shrews are not closely related to other shrews and are placed in an entirely different group, or order, called the Scandentia. Tree shrews are an important group of mammals because it is from their ancestors that the primates evolved. Unlike other shrews, they have large eyes and good vision, and look more like squirrels with long snouts. There are now only 16 species, all living in forested areas of eastern Asia.

Lesser tree shrew

Tupaia minor

Distribution: Malaysia, Sumatra and Borneo.
Habitat: Forests.
Food: Insects, small vertebrates, fruit, seeds and leaves.
Size: 14–23cm (5.5–9in); 100–300g (0.2–0.6lb).
Maturity: 3 months.
Breeding: 1–3 young.
Life span: 12 years.
Status: Not known.

This is the smallest of 11 species of tree shrew from the genus *Tupaia*, named after the Malay word for squirrel, *tupai*. These tree shrews do indeed resemble squirrels, with their long, bushy tails. Unlike most other species of small mammal, they are active throughout the day, though they also have short rests at regular intervals.

Their main predators are eagles, other birds of prey and snakes. Tree shrews are inquisitive, and are constantly searching in holes and crevices for food. Like squirrels, they hold their food in their front paws and eat while sitting on their haunches. Although they are called tree shrews and are capable climbers, they spend a lot of time on the ground or in the branches of bushes.

Tree shrews are highly territorial, with males defending an area of just over 1ha (2.5 acres), which covers the ranges of one or more females. They repel intruders of the same sex by chasing them and making loud squeals of aggression. Tree shrews also make other calls, which include chattering in response to disturbance, and clucking and whistling during courtship and mating.

Lesser tree shrews' fur colour ranges from a rusty-red colour through dark browns to grey. They have squirrel-like bushy tails and sharp claws for climbing trees.

Indian tree shrew

Anathana ellioti

Walter Elliot, an English civil servant based in Madras during the mid-19th century, was the first person to describe this species, hence its Latin name. However, very little has been learned about it since then, even though it seems to occur over much of southern India.

Unlike other species of tree shrew, this species is rarely found in close proximity to human dwellings. It has been seen in the dry deciduous forests on India's Deccan Plateau, and presumably has very similar habits to other tree shrews. It has also been found living in rocky upland habitats, without trees, at altitudes of 1,400m (4,600ft).

The Indian tree shrew spends a large part of the day searching for food and then returns to a hole among rocks before nightfall, where it shelters until dawn. These shrews appear to be solitary animals, almost always foraging alone and rarely, if ever, sharing shelters. However, occasionally small groups of three or four individuals have been observed playing together for short periods.

The Indian tree shrew, also sometimes called the Madras tree shrew, can be distinguished from other species of tree shrew living in India by its larger and more thickly haired ears.

Distribution: India, south of the Ganges River.
Habitat: Forests.
Food: Insects, earthworms and fruit.
Size: 17–20cm (7–8in); 160g (0.3lb).
Maturity: 3 months.
Breeding: 1–3 young born in related species.
Life span: Not known.
Status: Threatened.

PRIMITIVE PRIMATES

Although these primates have simpler social structures and smaller brains than the monkeys and apes, in other ways they are well adapted for their particular habitats. There are 57 species of primitive primate, the majority living in Madagascar, with some living in tropical Asia. They are all nocturnal tree-dwellers, and are characterized by their large, round eyes.

Tarsier

Tarsius bancanus

These little primates are nocturnal hunters, living in forests with thick ground-level vegetation. During the day, they sleep clinging to vertical branches, and slowly wake up as it gets dark. They move about through the branches like tree frogs, making leaps between branches. They have round pads on their fingertips and toes that help them grip on to surfaces.

When hunting at night, they scan for prey with their huge eyes, which measure 16mm (0.6in) across. Thanks to special vertebrae, they can turn their heads almost 180 degrees in either direction. When they spot prey, which may be an insect, a small bat or even a poisonous snake, they jump, catch their prey in their hands and kill it with a bite to the back of its neck.

Usually tarsiers live alone, but sometimes adult males and females live together. Tarsiers make a whole range of calls that serve to attract mates and warn off competitors. Females carry their babies in their mouths, or sometimes the babies ride on their mother's back. Baby tarsiers become independent very quickly, and after only 26 days they are able to hunt by themselves.

As well as their round eyes and long naked tails, tarsiers have very powerful back legs with extremely long anklebones – or tarsals – which give them their name.

Distribution: Southern Sumatra, Borneo and possibly Java.
Habitat: Forests, mangroves and scrub.
Food: Insects and small vertebrates.
Size: 9–16cm (3.6–6.4in); 80–165g (0.17–0.36lb).
Maturity: Not known.
Breeding: 1 young per year.
Life span: 13 years.
Status: Common but decreasing, partly due to habitat loss.

Slender loris

Loris tardigradus

The slender loris and its cousins the slow lorises are closely related to the bushbabies of Africa. Unlike the bushbabies, which have long fluffy tails, lorises lack tails. Like the tarsiers and other primitive primates, the slender loris is a tree-dwelling animal, mostly active at night. But unlike the tarsiers, the slender loris moves through the branches very slowly and deliberately, always supported by at least three of its limbs. The slender loris has opposable thumbs and big toes which it uses to grasp branches while climbing. When it spots prey, it cautiously stalks and, once it gets within range, grabs it with both hands.

Lorises usually live alone and, if forced to share space with others in captivity, they will fight one another. These animals make a variety of calls, including squeals and growls that indicate distress and disturbance, and whistles that are often heard when familiar males and females meet. The slender loris is threatened in India because its forest habitat is becoming increasingly fragmented, and because it is hunted for use in traditional medicines.

Distribution: Southern India and Sri Lanka.
Habitat: Mountain and lowland forests.
Food: Insects, young leaves, shoots, fruit, birds' eggs and small vertebrates.
Size: 18–26cm (7–10.5in); 85–348g (0.19–0.75lb).
Maturity: Females 10 months; males 18 months.
Breeding: 1–2 young born every 10 months.
Life span: 15 years or more in captivity.
Status: Vulnerable.

The slender loris differs from the slow loris in having a much more slender body and limbs. All lorises have thick woolly fur.

APES

There is only one species of great ape living outside of Africa – the orang-utan of Sumatra and Borneo. In contrast, all 11 species of lesser ape, or gibbon, live in tropical China, South-east Asia and the East Indies. Like the great apes, the gibbons are tailless and move in the treetops by swinging from branch to branch (brachiating), though with considerably more agility than their heavier, more terrestrial relatives.

Orang-utan

Pongo pygmaeus

Old adult male orang-utans can sometimes get very fat, and these have very round faces because of deposits of fat under their skin.

The orang-utan is unique among large primates because it is relatively solitary and lives in trees, whereas other large primates, such as humans, chimps, gorillas and baboons, live in social groups and spend most of their time on the ground. The name orang-utan means "forest people" in the Malay language.

Female orang-utans sometimes accompany each other while travelling and will eat together. Males, on the other hand, usually avoid one another. Orang-utans travel less than 1km (0.6 miles) per day through their home ranges, moving from branch to branch at a leisurely pace. These apes sleep in nests high in trees, made from leafy branches. Like the other great apes, orang-utans are intelligent animals, and have often been observed using tools. For example, they use large leaves as umbrellas to keep the rain off, and smaller leaves as pads to protect their hands and feet when moving through thorny vegetation.

Distribution: Sumatra and Borneo.
Habitat: Primary rainforests.
Food: Mostly fruit, but also other vegetable matter, insects, small vertebrates and birds' eggs.
Size: 1.25–1.5m (4–5ft); 30–90kg (66–200lb).
Maturity: 12–15 years.
Breeding: Single young born every 3–4 years.
Life span: 60 years.
Status: Vulnerable, due to habitat loss.

Siamang

Hylobates syndactylus

Siamangs are the largest of the lesser apes or gibbons. All species of gibbon live in trees, and are among the most agile animals in the world. The distance across the outstretched arms of a siamang can be up to 1.5m (5ft) – far greater than its standing height.

Gibbons use their long arms to swing, or brachiate, through the trees, and can move 3m (10ft) in a single swing. They do not tolerate other gibbons in their territories, and use loud calls which can be heard from several kilometres away to warn away intruders. Both male and female siamangs possess naked throat sacks that inflate with breath and resonate their calls. Sometimes male and female siamangs sing together in duets, the male making booms and loud screams and the female barking and booming.

To help grasp wide branches, gibbons have long thumbs, which attach at the wrist rather than the palm – like those of great apes and humans.

Distribution: Peninsular Malaysia and Sumatra.
Habitat: Rainforests up to 1,800m (6,000ft) altitude.
Food: Leaves, fruit, flowers, buds and insects.
Size: 75–90cm (30–36in); 8–13kg (17–29lb).
Maturity: 8–9 years.
Breeding: Single young born every 2–3 years.
Life span: 40 years.
Status: Threatened.

Lar gibbon

Hylobates lar

Lar gibbons live in small social groups consisting of an adult male and female, and up to four immature offspring. Like the siamangs, adult lar gibbons defend their territories from intruders by calling, often in duets with males and females alternating their calls. However, lar gibbons spend much more time calling than siamangs, making calls nearly every day.

Female lar gibbons are much less social than female siamangs and, although the males spend a lot of time showing apparent signs of affection such as embraces and grooming behaviour, the females rarely reciprocate. Males are probably more likely to lose their partners to competitors than siamangs, hence the greater amount of calling.

Lar gibbons sleep in the highest trees in the forest, usually those giants that emerge from the forest canopy. They almost never sleep in the same place from night to night, and this is thought to be a strategy to prevent predators, such as pythons and birds of prey, from learning their position.

All lar gibbons have white hands and feet, giving them their alternative name, the white-handed gibbon.

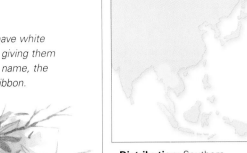

Distribution: Southern China, eastern and southern Burma, Thailand, Malaysian peninsula and northern Sumatra.
Habitat: Lowland tropical forest.
Food: Fruit, leaves and insects.
Size: 45–65cm (18–26in); 5–8kg (11–18lb).
Maturity: 8–9 years.
Breeding: 1 young born every 2–2.5 years.
Life span: 25 years.
Status: Endangered.

Kloss's gibbon (*Hylobates klossi*): 45–65cm (18–26in); 6kg (13lb)
This species is found only on a few islands in the Mentawai group, near western Sumatra. Due to its restricted range and low numbers, it is classified as vulnerable by the IUCN. Like other gibbons, this species defends territories in the forest, the males advertising their presence and strength to rivals by singing just before dawn.

Javan gibbon (*Hylobates moloch*): 45–65cm (18–26in); 4–9kg (9–20lb)
Javan gibbons are found only on Java, where they have lost over 95 per cent of their original forest habitat in recent decades. As a result, it is estimated that only 250 adults remain, making this the world's rarest gibbon.

White-browed gibbon (*Hylobates hoolock*): 45–65cm (18–26in); 4–9kg (9–20lb)
This gibbon has the most westerly distribution of all the gibbons, living as far west as Bangladesh. The white-browed gibbon is the only species of gibbon that can swim, while all other species avoid water. Although gibbons can walk upright, they almost never cross open ground on foot. This means that features such as rivers and open spaces will act as barriers to the spread of gibbon populations.

Crested gibbon

Hylobates concolor

This species used to be much more widespread than it is today; at one time it lived in much of southern China, and was found as far north as the Yellow River in eastern China. It has suffered considerably from the loss of its forest habitat to logging and agriculture. It has also been hunted for its meat, which is considered a delicacy by some Chinese people, and for its bones, which are prepared as a local medicine to treat rheumatism.

Wars in South-east Asia have helped throw this species into decline, so that it now numbers less than 2,500.

Like other gibbons, the crested variety moves around the treetops searching for fruit and tender leaves during the day. However, unlike any other species, crested gibbons sometimes live in polygamous groups, with one adult male sharing a territory with up to four adult females.

Distribution: South-eastern China, Hainan, north-western Laos and Vietnam.
Habitat: Tropical forest.
Food: Fruit, leaves and insects.
Size: 45–65cm (18–26in); 5–9kg (11–20lb).
Maturity: 8–9 years.
Breeding: Single young born every 2–3 years.
Life span: 25 years.
Status: Endangered.

Male crested gibbons are black, while females are golden or grey-brown in colour. Females are born almost white, and only attain their adult colour at between 2 and 4 years of age.

MONKEYS

Around half of the 96 species of Old World monkey live in south-central Asia, South-east Asia, Japan and Indonesia, where they are commonly known as langurs and macaques. Most Old World monkeys live in hot, tropical climates, but a few species, such as the Japanese macaque, can be found in very cold habitats at northerly latitudes.

Proboscis monkey

Nasalis larvatus

Scientists are not sure why proboscis monkeys, particularly older males, have evolved such strange noses, but they may be involved in attracting females or even used to radiate excess heat. Proboscis monkeys are very social and sometimes feed together in bands of up to 60 or more animals. Usually they live in smaller groups of 2–27 animals, consisting of a single dominant male, a harem of 1–9 females and their young.

Young females stay with the group, while young males move away to join bands of bachelor males. Proboscis monkeys rarely move more than 1–2km (0.6–1.2 miles) away from fresh water. They sleep close to rivers or in mangrove trees in coastal areas. These monkeys are among the best swimmers of all primates. They even have partially webbed feet, which help them to paddle or walk across soft mud.

Proboscis monkeys have peculiar protruding noses that become particularly long and bulbous in old males. The infants have blue faces.

Distribution: Borneo.
Habitat: Lowland rainforest and mangroves.
Food: Leaves, fruit, flowers and seeds.
Size: Females 53–61cm (21–24in), 7–11kg (15–24lb); males 66–76cm (26–30in), 16–22kg (35–48lb).
Maturity: Not known.
Breeding: Usually single young born at a time after a gestation of about 166 days.
Life span: 23 years.
Status: Vulnerable.

Crab-eating macaque

Macaca fascicularis

The crab-eating macaque is one of the most widespread of the 20 species of macaque, most of which live in parts of the Indian subcontinent, southern China, South-east Asia and Indonesia. Crab-eating macaques live in groups of around 30 individuals, which consist of adult males and females and their young. Crab-eating macaques live in ordered societies based on dominance hierarchies. Dominant individuals often force lower-ranking animals away from the best feeding and resting sites and give them few opportunities to mate. Indeed, low-ranking females take longer to reach sexual maturity than high-ranking females because they eat less food. Although crab-eating macaques prefer mangroves and forests around rivers, they have adapted to a range of habitats and sometimes live among people in towns. These monkeys sometimes raid crops and orchards, and they can be aggressive to humans.

Crab-eating macaques have tails that are longer than their bodies, giving them their alternative name, the long-tailed macaque.

Distribution: Malaysia, Indonesia and the Philippines.
Habitat: Forests, coastal mangroves and urban areas.
Food: Leaves, fruit, flowers, crustaceans, molluscs, insects, birds' eggs and small vertebrates.
Size: 40–47cm (16–19in); 3–7kg (6.6–15lb).
Maturity: Females 4 years; males 6 years.
Breeding: 1 young born every 2 years.
Life span: 35 years.
Status: Near threatened.

Hanuman langur

Semnopithecus entellus

These monkeys live in a wide variety of habitats, from hot tropical forests to cold mountain habitats, up to 4,000m (13,000ft) above sea level. Although they are principally tree-living animals, langurs are happy to live on the ground in habitats with few trees.

In areas with lots of food, hanuman langurs live in groups that may include a number of adult males as well as females and young, but in areas with less food, such as certain mountain habitats, there is only one adult male per group.

Like the crab-eating macaque, both sexes show a dominance hierarchy, though it is less pronounced in female hanumans than in female macaques. Males sometimes form groups of up to 30 individuals. Occasionally all-male groups attack male–female groups, and if they succeed in overthrowing the alpha male or males, the newcomers kill all the infants in the group. This makes the females come into oestrus quicker, so the new leaders can start to father their own offspring.

Distribution: Tibet, Nepal, Sikkim, northern Pakistan, Kashmir, India, Bangladesh and Sri Lanka.
Habitat: Tropical and temperate forests, savannah, farmland, alpine scrub and desert edges.
Food: Leaves, fruit and flowers.
Size: 41–78cm (16–31in); 5.5–23.5kg (12–52lb).
Maturity: Females 3–4 years; males 6–7 years.
Breeding: 1 or 2 young born every 2 years.
Life span: 25 years.
Status: Near threatened.

Hanuman langurs have dark faces and prominent brow ridges. They usually have grey, brown or buff-coloured upper-parts, and orange-white or yellow-white heads and chests. Their tails are longer than their bodies.

Pig-tailed langur (*Simias concolor*): 45–52cm (18–20in); 6.5–8kg (14–18lb)
This rare monkey is only found on a few islands in the Mentawai group off the south-west coast of Sumatra, and gets its name from its short naked tail with a tuft of fur on the tip. It lives in thick forest in hilly country. It is not as social as many other langur species and usually lives in pairs with 1–3 young.

Douc langur (*Pygathrix nemaeus*): 56–76cm (22–30in); 10kg (22lb)
This brightly coloured monkey is found in Cambodia, Vietnam and Laos. Like most monkeys it is very social, living in groups of 4–15 individuals consisting of mostly females, but including one or more adult males. This is one of the most brightly coloured species of monkey, with some varieties sporting bright yellow faces surrounded by pure white whiskers, and rich red-chestnut coloured legs.

Banded leaf monkey (*Presbytis femoralis*): 42–61cm (16–24in); 5–8kg (11–18lb)
This is the most widespread of eight species of leaf monkey from the same genus. It is found throughout the Malay peninsula, parts of Sumatra and Borneo and surrounding islands. Despite its broad range, this monkey is becoming rare. These leaf monkeys are characterized by crests of hair on the top of their heads. Sometimes they are forced out of their feeding areas by the smaller but more aggressive crab-eating macaque.

Snub-nosed monkey

Rhinopithecus roxellana

This is one of the few species of primate living exclusively outside the tropics, inhabiting the cold mountain forests on south-eastern slopes of the Tibetan plateau in China.

These monkeys are mostly tree-living, but will come to the ground occasionally to search for wild onions and to eat grass. In summer, when food is relatively abundant, up to 600 snub-nosed monkeys may live together, but as winter falls and food becomes more scarce, these large gatherings split up into smaller groups of 60–70 animals.

Snub-nosed monkeys have many vocalizations, the most common being a loud "ga-ga" call made when food is found.

There are probably fewer than 15,000 left in the wild because they have been hunted for their beautiful fur and for other parts that are used in medicines.

Distribution: Mountainous areas of the Tibetan plateau in south-western China.
Habitat: Mountain broadleaf and conifer forests.
Food: Leaves, buds, bark, grass and lichens.
Size: 57–76cm (22–30in); 12–21kg (26–46lb).
Maturity: Females 4–5 years; males 7 years.
Breeding: 1 or occasionally 2 young.
Life span: Not known.
Status: Vulnerable.

Although these monkeys have dull, grey-black fur on the top of their head, shoulders, back and tail, other parts of the body are covered in a rich, golden fur, giving them their other name, the golden monkey.

MONOTREMES

Some scientists do not consider the monotremes to be true mammals, but intermediate forms between mammals and reptiles. Indeed, their anatomy, digestion, reproductive organs and excretory systems have many similarities to those of reptiles, for example they lay eggs. However, monotremes also show mammalian characteristics: they have fur, they nourish their young with milk and they are warm-blooded.

Short-nosed echidna

Tachyglossus aculeatus

The short-nosed echidna is one of two species of echidna. It is found throughout Australia and parts of New Guinea. The other rarer species, the long-nosed echidna, lives only in the highlands of New Guinea. The short-nosed species regularly feasts on ants or termites, thus earning its common nickname, the spiny anteater.

Short-nosed echidnas live alone, foraging during both the day and the night. They locate their food by smell, and it is also thought that they may be able to pick up electric signals from prey animals using sensors in the tip of their snout. They are powerful diggers, using their clawed front paws to dig out prey or create burrows for shelter. The short-nosed echidna lays a single egg with a leathery shell into a pouch on its abdomen, and the young hatches soon afterwards. The baby is incubated in the pouch for 10–11 days, and finally leaves the pouch after around 55 days. The mother will continue to suckle her young even after it has left the pouch.

The short-nosed echidna has a long, hairless snout. The protective spines are enlarged hairs. They cover its back and sides, and are much thicker than those of the long-nosed echidna.

Distribution: Widespread throughout Australia, including Tasmania.
Habitat: Most habitats, from semi-desert to highlands.
Food: Ants, termites and earthworms.
Size: 30–45cm (12–18in); 2.5–7kg (5.5–15.5lb).
Maturity: Not known.
Breeding: Single young at a time, incubated in a pouch on the mother's abdomen.
Life span: 50 years.
Status: Low risk.

Platypus

Ornithorhynchus anatinus

Platypuses live in burrows in stream banks, which they excavate with their powerful front legs. There are two kinds of burrow: one for shelter and one for incubation. The incubation burrows can be up to 18m (60ft) long and rise 1–7m (3–23ft) above the waterline. As part of the courtship ritual, females carry bundles of wet leaves to their incubation chamber, at the ends of the burrow. Females then plug the tunnel with soil and lay 1–3 eggs in the incubation chamber. The eggs hatch after ten days and the young stay in the burrow, where the mothers keep them warm and suckle them. They finally emerge after four months.

Males possess spurs on their hind feet that are attached to venom glands, which produce poison strong enough to kill dogs. These are thought to be used in fighting.

This is surely one of the strangest living mammals, with a snout resembling a duck's bill, a tail like a beaver's, and clawed, webbed feet. The bill is thought to contain an organ for detecting the electric pulses produced by prey.

Distribution: Eastern and south-eastern Australia and Tasmania.
Habitat: Freshwater streams and lakes.
Food: Freshwater crustaceans, insect larvae, snails, tadpoles and small fish.
Size: 30–45cm (12–18in); 0.5–2kg (1.1–4.4lb).
Maturity: 2–3 years.
Breeding: 1–3 eggs laid at a time.
Life span: 13 years in the wild; up to 21 years in captivity.
Status: Fairly abundant, though may be declining.

MARSUPIALS

The marsupials are a group of mammals with a peculiar reproductive method. The young are very small and underdeveloped when born, and after making their way to their mother's nipples, they grow and develop in safety inside a pouch. Nearly 75 per cent of all marsupial species are found in and around Australia and New Guinea, with the rest being found in the Americas.

Squirrel glider

Petaurus norfolcensis

This is one of five closely related species of gliding possums, including the sugar glider, that live in Australia and New Guinea. These animals resemble flying squirrels, they are mostly nocturnal and have large eyes to help them see in low light. They are arboreal, and during the day they sleep in nests made from leaves in the hollows of trees.

Squirrel gliders have membranes of skin that stretch from their wrists to their ankles, which form gliding surfaces (aerofoils) when their legs and arms are spread out. The closely related but smaller sugar gliders can glide for distances of up to 45m (150ft), and the larger fluffy glider of the same genus can glide up to 114m (375ft).

Squirrel gliders live in social groups, usually consisting of one adult male, several adult females and their young. These animals communicate with a variety of calls, including chatters, nasal grunts and repetitive short grunts, which seem to serve as alarm calls and warnings to squirrel gliders from other groups. During the breeding season, females develop pouches in which the young stay for around six months. After this time, the pouches wither away.

Squirrel gliders have beautiful long tails, grey fur on top and pale fur underneath. They have attractive dark stripes from their noses all the way to their rumps, and stripes across their eyes.

Distribution: Eastern and south-eastern Australia.
Habitat: Open forests.
Food: Nectar, pollen, sap, insects and other invertebrates and small vertebrates.
Size: 12–30cm (5–12in); 200–260g (0.4–0.6lb).
Maturity: Females end of first year; males beginning of second year.
Breeding: 1 young, or occasionally twins, annually.
Life span: 12 years.
Status: Threatened.

Long-clawed marsupial mouse

Neophascogale lorentzii

This is one of over 50 species of marsupial mice, all of which live in Australia and New Guinea. All are carnivorous, taking any prey small enough to be overpowered. Very little is known about this species, probably because it lives in the remote and inaccessible wet forest habitat of the interior of New Guinea, at elevations of up to 3,000m (10,000ft). However, the long-clawed marsupial mouse is quite similar in appearance and habits to the better-known species living in Australia, such as the brush-tailed marsupial mouse.

Like the Australian species, the long-clawed marsupial mouse is nocturnal and arboreal, though it is also sometimes active during the day. It is highly territorial, with females occupying huge territories – about double the size of male territories. Females are very aggressive with one another and can usually dominate males. Daughters sometimes inherit parts of their mothers' territories when they become independent, but sons are always driven away to seek their own territories.

The long-clawed marsupial mouse resembles a tree shrew, with its long, thin muzzle and large, bushy tail. As its name suggests, it has particularly long claws on all four feet.

Distribution: Western and central mountains of New Guinea.
Habitat: Upland wet moss forests.
Food: Insects and perhaps other small animals.
Size: 17–23cm (7–9in); 110–230g (0.25–0.5lb).
Maturity: 8 months.
Breeding: 8 young.
Life span: 1–2 years.
Status: Not known.

Koala

Phascolarctos cinereus

Distribution: South-eastern Australia.
Habitat: A range of habitats, from coastal islands and tall eucalyptus forests to low woodlands inland.
Food: Eucalyptus leaves.
Size: 60–85cm (24–34in); 4–15kg (8.8–33lb).
Maturity: Sexual maturity at 2 years in females and males, but first breeding probably not until 5 years old in males.
Breeding: 1 young, or occasionally twins, born annually.
Life span: 20 years.
Status: Near threatened.

Koalas are placed in their own family, which shows how different they are from other marsupials. These animals live practically their whole lives in stands of eucalyptus trees that form the core of their territories. Eucalyptus leaves are by far the most important food for koalas. The animals even smell of eucalyptus oil. A problem with eucalyptus leaves is that they are very fibrous and difficult to digest, so koalas have cheek pouches and long intestines, and occasionally eat soil or gravel. All these traits are adaptations to help them break down the tough eucalyptus leaves.

Koalas are very solitary animals and, although the territories of males and females overlap, males will not tolerate intruders of the same sex. Resident koalas usually attack intruders savagely. Males mark out their territories with strong-smelling scent, and make loud calls to warn off other males.

During the breeding season, males try to guard as many female territories as they can. The single young or joey stays in its mother's pouch for seven months, then rides on its mother's back until the next season's joey matures. Koalas used to be very common in south-eastern Australia, but a mixture of hunting for fur in the early 20th century, habitat loss and severe forest fires have reduced their numbers.

Koalas have dense grey fur, which is paler on their undersides. They have two opposable thumbs and three fingers on each of their hands, which helps them to grip strongly on to branches.

Common wombat

Vombatus ursinus

This is the most common of three species of wombat, all of which live in Australia. Wombats resemble small bears and have large heads, short, stocky bodies and short, powerful legs. The common wombat can be distinguished from the other two species by its hairless nose and rounded ears. Unlike bears, wombats are not predators, and tend to be shy and timid animals. In fact, they make good pets, being playful and affectionate.

In the wild, wombats dig out large burrows that may be up to 30m (100ft) long. Although there is only one entrance, a burrow may branch off into different chambers. The wombat makes a nest out of leaves or bark in one of the chambers.

Wombats are usually nocturnal, but sometimes they sunbathe in specially prepared spots, close to the entrances of their burrows. Common wombats live in large home ranges and, although they tend to be solitary, neighbouring territories overlap considerably. Although wombats are sometimes aggressive towards each other in captivity, some observations in the wild suggest that wombats sometimes make visits to one another's burrows.

Distribution: Originally found throughout south-eastern Australia and Tasmania.
Habitat: Hilly or mountainous coastal country, creeks and gullies.
Food: Grass, roots of shrubs and trees, and fungi.
Size: 0.7–1.2m (28–48in); 15–35kg (11–77lb).
Maturity: 2 years.
Breeding: 1 or occasionally 2 young born in late autumn after 20 days' gestation.
Life span: 25 years.
Status: Common.

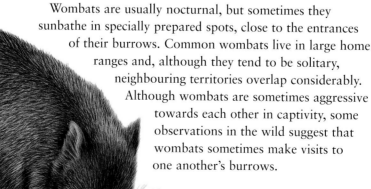

Wombats resemble small, chubby bears, having compact bodies covered with a dense coat of grey-brown hair. Evidence from fossils suggests that there were once hippo-sized wombats.

Brush-tailed possum

Trichosurus vulpecula

Large numbers of Australasian marsupials are commonly referred to as possums, which belong to several families. These are not to be confused with opossums, which are marsupials from an entirely separate group, living in the Americas.

This possum is the most common of three possum species, all known as brush-tailed possums due to their long, furry prehensile tails. It is closely related to the cuscuses of Sulawesi, New Guinea and surrounding islands.

The brushed-tailed possum is nocturnal and arboreal. It is a very adaptable animal, and can live in semi-desert areas in the Australian interior by sheltering in eucalyptus trees and along rivers. It even lives in large numbers in most Australian cities, hiding under the roofs of houses.

Possums generally live alone and defend well-defined territories marked with scent. Encounters often result in fights. When these animals are disturbed, they rear up to their full height on their back legs, spread their forelimbs and make piercing screams. They are fierce fighters.

Brush-tailed possums come in a variety of colours, including grey, brown, black, white and cream. Males tend to have red-brown coloration on their shoulders.

Distribution: Most of Australia, including Tasmania. Introduced into New Zealand.
Habitat: Forests, rocky areas, semi-deserts with scattered eucalyptus trees and suburban habitats.
Food: Shoots, leaves, flowers, fruit, seeds, insects and occasionally young birds.
Size: 32–58cm (12–23in); 1.3–5kg (2.8–11lb).
Maturity: Females 9–12 months; males 24 months.
Life span: 13 years.
Status: Common.

Numbat (*Myrmecobius fasciatus*): 17–27cm (7–11in); 300–600g (0.6–1.3lb)
This marsupial from southern parts of Australia is also known as the banded anteater and, as the name suggests, it specializes in eating ants and termites. Like other anteaters, it has a long snout, a sticky tongue, which can extend at least 10cm (4in), and powerful claws for ripping open rotting logs to expose termite nests.

Marsupial mole (*Notoryctes typhlops*): 9–18cm (3.5–7in); 40–70g (0.08–0.15lb)
This unusual species is not related to other moles, and is only distantly related to other marsupials. It is white or cream in colour, but otherwise resembles a placental mole. It spends a lot of time underground and has long, spade-like claws on its forelimbs for digging, and tiny eyes hidden under its skin. This is a good example of "convergent evolution" between marsupial mammals and placental mammals – having evolved similar adaptations for the same lifestyle.

Native cat or **quoll** (*Dasyurus geoffroi*): 29–65cm (11–26in); 500g (1.1lb)
This is one of six species, all from the same genus, that are called native cats or quolls. They are not related to the true cats, but get their name from their somewhat cat-like appearance and their predatory behaviour. All species of native cat have coats with white spots. They are unpopular with farmers because they occasionally raid chicken coops. As a result of habitat loss, persecution and possibly competition and predation by introduced red foxes, all species of native cat are declining.

Tasmanian devil

Sarcophilus harrisii

The Tasmanian devil comes from a family of carnivorous marsupials that includes marsupial mice and native cats, or quolls, which used to live all over Australia. They are now restricted to the island of Tasmania, which was never colonized by dingoes.

The Tasmanian devil is nocturnal and spends a lot of its time snuffling over the ground trying to pick up the scent of food. It is a very efficient scavenger, and uses its powerful jaws to crush bones and chew up tough skin.

During the day, Tasmanian devils take refuge in nests of bark, grass or leaves inside hollow logs, old wombat burrows or other sheltered spots. The ferocity of Tasmanian devils has been greatly exaggerated, and although they will sometimes fight savagely amongst themselves when feeding, they are apparently docile and safe to handle.

Distribution: Originally found over much of Australia, but now only occurs in Tasmania.
Habitat: Coastal heath and forest.
Food: Carrion, small invertebrates and vertebrates.
Size: 52–80cm (20–32in); females 4.1–8.1kg (9–17lb); males 5.5–12.8kg (12–28lb).
Maturity: 2 years.
Breeding: Litter of 1–4 young.
Life span: 8 years.
Status: Common.

These stocky little carnivores look like tiny bears, but have long tails. They usually have black or dark brown fur, with white markings on their throat, rump and sides.

Musky rat kangaroo

Hypsiprymnodon moschatus

Musky rat kangaroos have dense rich brown or rusty grey fur that is lighter on their undersides. Compared to other rat kangaroos, their back legs are less well developed. These animals carry seeds away from the area where the fruit falls and bury them for eating later when food is scarce. Some seeds germinate, and produce trees for future seed supplies.

This is the smallest of ten species of rat kangaroo, all of which live in Australia. Most rat kangaroos have short fur on their long tails, but some have tails that are hairless and scaly. Like kangaroos, these animals have large feet and well-developed hind legs. However, unlike other species in the family, this rat kangaroo usually moves about on all fours rather than hopping kangaroo-fashion.

Both the males and females emit musky scents, probably to attract mates. Although they usually live alone, they do not seem to defend territories. Unlike all other rat kangaroos, they are most active during the day, when they forage on the forest floor, looking for insects and worms by turning over forest debris and rummaging in the leaf litter with their forepaws. Sometimes they take a break from foraging and stretch out to sunbathe in bright spots in the forest. At night, musky rat kangaroos sleep in nests between the buttress roots of trees or in tangles of vines. The range of these animals is contracting because of forest clearance for agriculture.

Distribution: North-eastern Queensland, Australia.
Habitat: Dense vegetation in tropical rainforest.
Food: Insects, worms, roots, seeds, palm fruit, fungi, flowers, twigs, leaves, lichen and bark.
Size: 21–34cm (8–14in); 337–680g (0.75–1.5lb).
Maturity: Between 18 and 21 months.
Breeding: Usually 2 young born at a time.
Life span: Not known.

Red kangaroo

Macropus rufus

Distribution: Most of Australia.
Habitat: Grassland and savannah with some cover.
Food: Grass.
Size: 0.8–1.6m (32–64in); 20–90kg (44–180lb).
Maturity: 2 years.
Breeding: 1 or 2 born when conditions are favourable.
Life span: 27 years.
Status: Common.

Like other members of the family, red kangaroos have large bodies with a relatively small head, large ears, well-developed hind legs, huge feet and a long muscular tail. Males are usually a rich red-brown colour, while females are blue-grey.

The red kangaroo is the largest of the 61 species comprising the kangaroo and wallaby family, and stands up to 1.8m (6ft) tall on its hind legs. This marsupial mammal has a unique way of travelling. When grazing, it moves around slowly by supporting its body on its forelegs and tail and swinging its back legs forwards. When moving fast however, it hops on its powerful back legs and can make single leaps of more than 9m (30ft).

This is a very efficient way of moving across rough terrain, and kangaroos bounce along at speeds of up to 48kph (30mph) on their springy back legs, using much less effort than a running placental mammal of similar size.

Kangaroos live in dry conditions and can live for long periods without water. They are able to survive on low-quality vegetation found in dry habitats, thanks to a digestion process that uses gut bacteria to help break down tough plant material. However, unlike other large species of kangaroos, reds usually move in search of better feeding conditions during droughts, sometimes travelling more than 200km (125 miles). Red kangaroos are not territorial, but males will fight one another by boxing with their arms and kicking with their back legs, to win control of groups of females during the breeding season.

Doria's tree kangaroo

Dendrolagus dorianus

Distribution: Western, central and south-eastern New Guinea.
Habitat: Mountainous rainforests.
Food: Mostly leaves and fruit.
Size: 52–81cm (20–32in); up to 20kg (40lb).
Maturity: Not known.
Breeding: 1 young born at a time.
Life span: 20 years.
Status: Vulnerable.

There are ten species of closely related tree kangaroo, all of which live in New Guinea and on a few surrounding islands, or in tropical northern Queensland in Australia. This species is the largest of the tree kangaroos and, like others in the group, it has a thick tail that can be longer than the body. Unlike most ground-living kangaroos, which have well developed back legs, the front and back limbs of tree kangaroos are almost the same size.

As their name asserts, tree kangaroos live in trees, but they frequently come to the ground to search for food and to move to new trees. Indeed, this species actually spends most of its time on the ground, while smaller species may spend as much as 98 per cent of their time in the branches of trees.

Tree kangaroos are very agile, and can jump between trees, dropping as much as 9m (30ft) from one branch to another. They are able to jump to the ground from as high up as 18m (60ft). Most species of tree kangaroo live alone, but related females of this species sometimes gang together to drive away unfamiliar males. Most species of tree kangaroo are declining because of increased hunting and habitat loss.

This species of tree kangaroo has fur in various shades of brown, but other species are among the most colourful of all marsupials, with bright yellow markings and different shades of red.

Large Celebes cuscus (*Ailurops ursinus*): 0.5–0.6m (20–24in); 7–10kg (15–22lb
This animal only lives on the Indonesian island of Sulawesi (Celebes) and belongs to the same family as the Australian possums. It has a stocky body with thick, dark-coloured fur above and pale fur below. Unlike the other two species of cuscus living on Sulawesi, this species lives in trees and has a long prehensile tail to help it climb.

Rock wallaby (*Petrogale lateralis*): 50–80cm (20–32in); 3–9kg (6.6–18lb)
There are 14 species of rock wallaby, all from Australia and all characterized by long, powerful hind legs, long tails and large ears. These animals live in dry rocky habitats, close to cover. They are incredibly agile on cliffs and rocky outcrops. They can make leaps of 4m (13ft) and easily bound up the steepest slopes.

Mountain pygmy possum (*Burramys parvus*): 10–13cm (4–5.2in); 30–60g (0.06–0.13lb)
This species was only known from fossils until 1966, when a live specimen was caught in a ski hut on Mount Hotham in south-western Australia. It is now known that around 1,000 individuals survive high up in the mountains, and pass the cold winters in hibernation. In spring and summer, the animals forage for seeds, fruit, insects and worms at night, sometimes climbing into bushes with the help of their prehensile tails.

Bilby

Macrotis lagotis

Bilbies, or rabbit-eared bandicoots, belong to a distinct group of marsupials with 22 species, all of which live in Australia, Tasmania, New Guinea and on some of the surrounding islands. They are characterized by their long, pointed muzzles, and have hind legs that are larger and more well developed than their front legs, though not to the same degree as the kangaroos.

There used to be two species of bilby, but the smaller of the two has not been seen since 1931 and is probably extinct. Unlike other species of bandicoot, bilbies are powerful diggers and excavate spiral-shaped burrows.

Bilbies are nocturnal, sleeping in their burrows during the day. They live alone or in small colonies, usually with one adult male, several females and a number of young. Bilbies have the shortest gestation of any mammal – 14 days. Once born, the babies spend around 80 days in their mother's pouch, followed by a further two weeks in the burrow.

Distribution: Originally found over most of temperate and arid Australia; now restricted to pockets in Western Australia, Northern Territory and south-western Queensland.
Habitat: Dry habitats, including dry woodland, scrub, savannah and semi-desert.
Food: Insects, small vertebrates and some vegetable material.
Size: 29–55cm (11–22in); 0.6–2.5kg (1.3–5.5lb).
Maturity: Females 5 months; males 12 months.
Breeding: 1 or 2 young.
Life span: 7 years.
Status: Endangered.

Males are usually bigger than females, but both have very long finely furred ears. Bilbies have long grey, black and white tails with a crest of long hair on the end.

INSECTIVORES

Insectivores are perhaps the oldest mammal group, dating back 135 million years. Today, the group includes the smallest mammals in the world. All insectivores have five digits on each limb and they usually have small eyes, long snouts and primitive teeth. As their collective name suggests, insectivores commonly feed on insects and other small invertebrates.

Moonrat

Echinosorex gymnura

Moonrats are related to hedgehogs and are one of the largest insectivores. Females tend to be larger than males, but both sexes are otherwise similar in appearance, with narrow bodies, long snouts and coarse black fur with white markings on their heads. Their exceptionally narrow bodies allow the creatures to search for prey in tight spaces. Moonrats are nocturnal, and during the day they sleep in protected spots, such as hollow logs, under tree roots or in holes. At night they search through the leaf litter for small prey.

Moonrats usually live close to water and sometimes swim in streams in search of fish, frogs, crustaceans and other aquatic prey. They sometimes carry parts of their prey to their resting sites, to be eaten later. Moonrats are solitary animals and except during mating, they do not tolerate each other's presence. They mark out their territories with strong-smelling scents to warn off intruders. Moonrats often respond to encounters by making hissing noises and low roars.

Moonrats have long, hairless and scaly tails up to 30cm (12in) long.

Distribution: South-east Asia.
Habitat: Lowland forests, plantations, mangroves and agricultural land.
Food: Worms and other leaf litter invertebrates, fish, amphibians and aquatic invertebrates.
Size: 26–46cm (13–23in); 0.5–2kg (1.1–4.4lb).
Maturity: 1 year.
Breeding: 2 litters of twins born every year.
Life span: 4 years.
Status: Common.

Indian hedgehog

Paraechinus micropus

The Indian or desert hedgehog is similar in appearance to the common European hedgehog, except that it tends to be smaller in size. Indian hedgehogs come in different colours; some have banded spines of dark brown, with black and white or with yellow, and there is also an unusually large proportion of black and white individuals.

As their alternative name suggests, these animals are well adapted to living in dry desert conditions. They escape the heat of the day by lying up in burrows about 1–2m (3–6ft) deep, which they dig themselves. When the temperatures drop at night, they come out and search the desert floor for prey. If they are alarmed or are chasing frogs, these hedgehogs can move surprisingly fast considering their short legs, reaching speeds exceeding 2kph (1.25mph).

They often bring food back to their burrows for later use. When water or food is scarce, Indian hedgehogs may stay sleeping in their burrows for long periods to conserve energy and water. These hedgehogs stay in the same places all year round and live alone.

Some Indian hedgehogs have a banded appearance with a brown muzzle and a white forehead and sides. Unlike European hedgehogs, which have smooth spines, they have rough spines.

Distribution: Pakistan and India.
Habitat: Deserts and other dry habitats.
Food: Insects, small vertebrates and birds' eggs.
Size: 14–27cm (5.6–13.5in); up to 435g (1lb).
Maturity: 1 year.
Breeding: 1–6 young born per litter.
Life span: 7 years.
Status: Lower risk.

Tibetan water shrew

Nectogale elegans

There are several species of shrew that are at home in water. All of these have silky fur that repels water, long tails and fringes of stiff hairs along the edges of their feet, toes, fingers and tails that help them swim. In smaller species, these hairs allow the shrews to run across the surface of the water for short distances, supported by the surface tension.

The Tibetan water shrew is the only species of shrew to have webbed feet. It also has disc-like pads that may help it keep its footing on slippery wet stones. This shrew lives in a burrow dug in a stream bank.

Most aquatic shrews forage by making repeated dives in the same spot, each dive lasting usually less than 20 seconds. After each dive, they shake their fur dry. If no prey is found, they move along the stream 1m (3ft) or so and dive in a new position. The Tibetan water shrew has sharp teeth that seem to be specialized for catching fish, which are its main prey.

The Tibetan water shrew has a long dark coloured tail with several fringes of short, stiff white hairs that shine with rainbow iridescence when wet.

Distribution: Tibet, south-central China, Nepal and northern Burma.
Habitat: Mountain forest streams.
Food: Small fish and aquatic invertebrates.
Size: 9–13cm (3.5–5in); 25–45g (0.05–0.09lb).
Maturity: 6 months.
Breeding: Not known.
Life span: Not known.
Status: Common.

Asian musk shrew (*Suncus murinus*): 7.5–12cm (3–4.8in); 10–32g (0.02–0.07lb)
This species lives in large numbers in and around human settlements, and is very common in houses throughout its range. It originated in India, but has been spread by people, right the way down the coast of East Africa and all the way east across Asia to Indonesia and a number of oceanic islands, including Mauritius and Guam, where it threatens small native reptiles.

Mole shrew (*Anourosorex squamipes*): 8.5–11cm (3.4–4.5in); 14–25g (0.03–0.05lb)
This strange shrew lives in large parts of China and South-east Asia and, although it is in the shrew family, it looks and behaves much more like a mole. It lives in mountain forests up to 3,100m (10,000ft) above sea level, and spends its time underground, burrowing among the roots of plants, searching for insects and earthworms that form its diet.

Asian mole

Euroscaptor micrura

There are six species of mole living in South-east Asia, southern China and Japan. Like the more familiar European moles, they spend much of their time underground and their bodies have a number of special adaptations to suit their subterranean lifestyle.

Members of the mole family, the *Talpidae*, generally dig tunnels in soil where earthworms are in abundance, along with other prey, and where they are safe from marauding predators. They possess large, powerful front paws on short, stocky forelimbs, which enable them to dig rapidly through even hard ground while using their shovel-like snouts to push loose soil aside.

The Asian moles have the largest front paws in relation to body size of all the moles. In the darkness of their tunnels they have no need of sight and so their eyes are tiny, and probably only able to detect light and dark. However, like many other moles they are very sensitive to ground vibrations, and can use them to find moving prey.

Two species of Asian mole have become rare due to habitat destruction, and one of these species, from a small region of Vietnam, is considered to be critically endangered.

Like other moles, Asian moles have short, velvety hair that will lie in the direction in which it is brushed. This makes it easy for moles to move both forward and backwards through tight tunnels.

Distribution: Nepal, Sikkim, Assam, northern Burma and southern China.
Habitat: Forests with deep soils in mountainous regions.
Food: Insects, earthworms and other soil invertebrates.
Size: 10–16cm (4–6.5in); 29g (0.06lb).
Maturity: Not known.
Breeding: Litters of 2–5 young.
Life span: Not known.
Status: Lower risk.

SEA MAMMALS

There are two main groups of sea mammals: the cetaceans (whales and dolphins) and the pinnipeds (seals, sea lions and walruses). The cetaceans evolved from a primitive group of hoofed mammals and became the masters of the oceans. The pinnipeds evolved from a carnivore ancestor and, although they can stay submerged for over an hour, they still have to come on to land to give birth.

Ganges river dolphin

Platanista gangetica

This species has a long, slender snout and a very low, ridge-like dorsal fin. The side, or pectoral, fins of this species have very square back edges.

The large majority of dolphins live in saltwater marine habitats. However, there are at least seven species of dolphin that regularly visit or permanently live in the freshwater habitats of large rivers, including the Ganges in India, the Yangtze in China and the Amazon in South America.

Because rivers often carry a lot of cloudy sediment, especially during times of heavy rain, visibility can be very poor. Indeed, these dolphins have very small deep-set eyes that lack lenses and are probably only useful for detecting light and dark. Ganges river dolphins emit clicks frequently and rely strongly on echolocation to find their way around. These dolphins also hunt for prey using echolocation and by probing in the mud on the river bottom using their sensitive snouts.

These dolphins may inhabit fast-flowing rivers, where the flow can be violent, especially during times of flood. This means that they have to be able to be swim 24 hours a day to avoid being washed away and injured.

Distribution: Ganges, Brahmaputra, Meghna Rivers and Karnaphuli River of India, Nepal and Bangladesh.
Habitat: River habitats, from the foothills of the Himalayas to tidal limits.
Food: Fish, shrimp and other bottom-dwelling invertebrates.
Size: 2–3m (6.5–10ft); 51–89kg (1122–195lb).
Maturity: 10 years.
Breeding: Single young born at a time.
Life span: 30 years.
Status: Endangered.

Hector's dolphin

Cephalorhynchus hectori

Distribution: New Zealand.
Habitat: Muddy waters.
Food: Fish, crustaceans, squid and invertebrates.
Size: 1.1–1.8m (3.5–6ft); 26–86kg (57–190lb).
Maturity: 6–9 years.
Breeding: Every 2–3 years.
Life span: 20 years.
Status: Vulnerable.

Hector's dolphins are among the rarest species of marine dolphin, with probably less than 5,000 individuals remaining. These dolphins only live in shallow areas around the coast of New Zealand.

Hector's dolphins live in groups of 2–8 individuals, though they may occasionally come together in aggregations of as many as 50 individuals. They hunt for fish and squid from the surface to the sea floor, and sometimes they follow the nets of trawlers in search of stray fish.

The critically endangered population around New Zealand's North Island has only 100 adults, but is unlikely to be helped by individuals from populations around the South Island because of very low migration rates. The northern population is almost certainly doomed to extinction.

Like other species in the same genus, Hector's dolphins are characterized by distinctive black and white markings. Their heads, pectoral and dorsal fins and tails are black and their flanks are grey, but their undersides are white. There are also two characteristic fingers of white going from their bellies and along their sides towards their tails – a pattern unique to this species.

Sperm whale

Physeter catodon

The sperm whale is supremely well adapted to life in the deep oceans. These are the largest hunting predators in the world, with teeth up to 20cm (8in) long and the largest brains of any mammal, weighing over 9kg (20lb). They prefer areas of ocean with cold upwellings at least 1km (3300ft) deep where squid – their favourite food – are most abundant. Sperm whales can dive to incredible depths to hunt, occasionally up to 2.5km (1.5 miles). They are social animals, and they live in groups of between 20 and 40 females, juveniles and young. Sperm whales have been hunted for their oil since the mid-18th century, and after serious population declines between the 1950s and 80s, this species is now protected.

The box-like head of the sperm whale contains the spermaceti organ, which is filled with the fine oil so valued by whalers. The purpose of this organ is unclear.

Distribution: Ranges throughout oceans and seas worldwide.
Habitat: Deep oceans.
Food: Mostly squid, including giant deep-sea squid, but also several species of fish and shark.
Size: 12–20m (40–65ft); 12,000–50,000kg (12–50 tonnes).
Maturity: Females 7–13 years; males 25 years.
Breeding: 1 calf born every 5–7 years.
Life span: 77 years.
Status: Vulnerable.

Leopard seal (*Hydrurga leptonyx*): 2.4–3.4m (8–11.25ft); 200–591kg (440–1,300lb)
These predatory seals live all around the Antarctic and occasionally they can be found in temperate waters around New Zealand, southern Australia and Argentina. They have large, sleek bodies, almost reptile-like heads and long canine teeth. Leopard seals eat krill, which they filter from the water using their cheek teeth. They are fearsome predators too, and also eat other seals and penguins.

Australian sea lion (*Neophoca cinerea*): 2–2.5m (6.5–8.25ft); up to 300kg (660lb)
This species lives around the coast of southern Australia, from Shark Bay in Western Australia to the south-eastern edge of Australia. The Australian sea lion eats fish, squid and crustaceans. It forms breeding colonies of around 100 individuals on small offshore islands, where dominant males defend small territories centred on females. As well as being strong swimmers, these sea lions are surprisingly adept on land, and can even climb steep cliffs.

Chinese river dolphin (*Lipotes vexillifer*): 1.4–2.5m (4.75–8.25ft); 42–167kg (92–334lb)
This freshwater dolphin lives in the Yangtze River in China from its estuary to around 1,900km (1,200 miles) upriver. Like the Ganges river dolphin, it has small eyes, a long, thin snout and a reduced dorsal fin. This species has suffered extensively from boat traffic strikes, hunting and the decline of its prey brought about by the construction of dams and pollution. It is probably the most endangered of all the dolphins.

Baikal seal

Phoca sibirica

This is the only one of 33 species of seal and sea lion that lives exclusively in fresh water. It has one of the most restricted distributions of all the pinnipeds, only being found in Lake Baikal in south-central Siberia.

In winter the lake is entirely covered in a thick layer of ice. During this period, the seals spend most of their time under the ice, coming up to breathe through access holes which they keep open by scratching with the claws on their front flippers and by abrading the edges with their teeth and heads.

Baikal seals dive for about 25 minutes when foraging for fish, but can remain submerged for an hour if frightened. During the breeding season in May, successful males gain access to harems of several females. Mating occurs underwater, and the pups are born in late winter or early spring. Females construct a chamber in the snow, where they give birth to and suckle their young.

The long whiskers are very sensitive to touch and are probably very important for locating prey in the dark conditions under the winter ice.

Distribution: Lake Baikal.
Habitat: Fresh water.
Food: Fish and aquatic invertebrates.
Size: 1.1–1.4m (3.5–4.75ft); 50–130kg (110–260lb).
Maturity: 6–7 years.
Breeding: Usually 1 pup, but occasionally twins.
Life span: 55 years.
Status: Near threatened.

INDEX

ACKNOWLEDGEMENTS
The publisher would like
to thank the following for
permission to reproduce their
photographs in this book.

Key: l=left, r=right, t=top,
m=middle, b=bottom

NHPA: 8bl, 16tr, 16b, 18t, 18b,
19t, 21t, 21br, 24br, 26t, 29b,
30t, 31t, 31b.
Tim Ellerby: 2, 23tl, 24t, 24bl,
27b, 28l, 31m, 34r.